SIXTH EDITION

Geology

ROBERT J. FOSTER

Macmillan Publishing Company
New York
Collier Macmillan Canada, Inc.
Toronto
Maxwell Macmillan International Publishing Group
New York Oxford Singapore Sydney

Senior Editor: Robert McConnin
Production Buyer: Janice E. Wagner
Production Editor: Rex Davidson
Art Coordinator: Lorraine Woost
Photo Editor: Gail Meese
Cover Designer: Brian Deep

Cover Photo © James Cowlin.

This book was set in Meridien.

Printed in the United States of America

Macmillan Publishing Company
866 Third Avenue, New York, New York 10022

Collier Macmillan Canada, Inc.

Library of Congress Cataloguing-in-Publication Data
Foster, Robert J. (Robert John), 1929 Apr. 19-
Geology/Robert J. Foster.—6th ed.
p. cm.
Includes bibliographical references and index.
1. Geology. I. Title.
QE28.F76 1991 90-43726
550—dc20 CIP

Printing: 2 3 4 5 6 7 8 9 Year: 1 2 3 4

Contents

Introduction

Geology is the study of the earth. That we live on the earth is reason enough to study it. The more we know about our planet, especially its environment and resources, the better we can understand, use, and appreciate it. To us, the earth is the most important body in the universe.

In the broader view, however, the importance of the earth shrinks. Only one other planetlike object has been observed orbiting a star. Several other stars are known to have faint, unseen companions, but it is not known whether these companions are very faint stars or massive planets revolving about the observed star. Thus, the earth appears to be an average planet. However, at least in the solar system, the earth is unique in having abundant water, exemplified by the oceans, and an atmosphere that can support life. The earth's surface temperature, controlled largely by its distance from the sun, makes these features possible, and these features, in turn, make life possible on the earth. The development and history of life are important aspects of historical geology. The space program also has revealed that the earth is unique among the earthlike planets so far studied in that it has a magnetic field. The earth's magnetic field, for reasons that will be discussed later, is believed to be caused by a liquid iron core that also may store the energy that causes the formation of such surface features as mountain ranges. These features also appear to be unique to the earth and are central to the processes of erosion and deformation of the earth's surface that are the main aspects of physical geology. Thus, geology is in large part the study of the consequences of the earth's unique features.

Geology has contributed a great deal to civilization, both intellectually and economically. Among the great concepts gained from geological studies are an understanding of the great age of the earth and the development of an absolute time scale. Geology differs from most other sciences in that it is concerned with **absolute time.** Time appears in the equations of physics and chemistry, but these sciences are generally concerned with rates of change, and the time is relative, not absolute. Geologic time extends back almost 5000 million years to when the earth formed. Thus, geology is concerned with immense lengths of time when measured against human experience. It is difficult to comprehend the lengths of time involved in geologic processes, but this must be done to appreciate geology fully.

Another important point learned from geology is that constant change, both biological and physical, has been and is occurring on the earth.

The economic contributions of geology to civilization show that, in many ways, geology also is a very practical science. Geologic knowledge is used to locate and to exploit our mineral resources. Except for water and soil, all mineral resources, such as sand and gravel, petroleum, coal, and metals, are nonrenewable. Once mined, they are gone, and new deposits must be found. Geologists have discovered the deposits of metal and energy-producing minerals on which our civilization is based. We take these things for granted now; but a hundred years ago, when the American West was opening up and the industrial revolution was in progress, these mineral deposits were being discovered at a rapid rate and geologists were the most influential scientists of the day.

Today the study of mineral resources is still important, but geologists are also concerned with other economic problems, such as urbanization. The development of large cities has resulted in the building of large structures such as tall buildings and dams. Geology helps in designing foundations for these structures. Examples are commonplace of both large and small structures that have failed through neglect of simple geologic principles, easily understood by elementary students. Dams fail because they are built near active faults or on porous foundations. During their first rainy season, new highways are washed out or blocked by landslides. Homes built on hillsides are destroyed by landslides and mudflows. Geologists have also recognized the need for earthquake-resistant structures in some areas and have helped in their design.

Geology is based mainly on observations and seeks to determine the history of the earth by explaining these observations logically, using other sciences such as physics, chemistry, and biology. Only a small part of geology can be approached experimentally. For example, although the use of fossils to date or establish contemporaneity of rock strata is based on the simple, basic principle that life has changed during the history of the earth, this principle could not be established experimentally; it was the result of careful observations and analyses over a long period of time by many people of varied backgrounds.

Geologic problems are many, diverse, and complex; almost all must be approached indirectly, and in some cases, different approaches to the same problem lead to conflicting theories. It is generally difficult to test a theory rigorously for several reasons. The scale of most problems prevents laboratory study—for example, one cannot bring a volcano into the laboratory, although some facets of volcanoes can be studied indoors. It is also difficult to simulate geologic time in an experiment. All of this means that geology lacks exactness and that our ideas change as new data become available. This is not a basic weakness of geology as a science, but means only that much more remains to be discovered; this is a measure of the challenge of geology.

Reasoning ability and a broad background in all branches of science are the main tools of geologists. Geologists use the method of *multiple working hypotheses* to test their theories and to attempt to arrive at the best-reasoned theory. This thought process requires as many hypotheses as possible and the ability to devise ways to test each one. Not always is it possible to arrive at a unique solution—but this is the goal. In the sense that geologists use observation, attention to details, and reasoning, their methods are similar to those of fictional detectives.

THE MOBILE EARTH

The earth is a dynamic planet whose surface is constantly undergoing change. Given the length of geologic time, one can comprehend that the hills and mountains are not eternal features, but that they eventually will be worn flat by erosion. Earthquakes and volcanoes reveal a restless interior. Deformed rocks that have been bent into folds and broken also bear mute testimony to the power of the earth's internal energy. An overview of how some of these changes occur will help in understanding the rest of this book.

The idea that the continents have moved around on the earth's surface is a very old one. Almost every school child notices that the continents fit together like pieces of a jigsaw puzzle. The fit is not perfect, but erosion and deformation could have distorted the pieces. The fit of the continents involves more than just shape; the rock layers and types also match. The ages of the rocks and their structures also are similar.

Therefore, the evidence is strong that at least the southern continents were once a single continent and that this supercontinent broke up and the pieces drifted apart. Although initially this concept was widely accepted by geologists in the southern continents, it was not believed by most American and European geologists. Their lack of acceptance was due in part to the fact that rocks in the northern continents do not match so clearly as those in the south, but mostly because no mechanism was known that could cause continental drift.

We now believe that the ocean floors themselves are moving, carrying the continents like a conveyor belt. All of the ocean bottoms have ridges or rises, called **mid-ocean ridges** (also called spreading centers and divergent plate borders). These are areas of volcanic activity and shallow earthquakes. New ocean floor is created at the ridges by the volcanic activity (Fig. I.1). This is shown by the age and thickness of the rocks on the ocean floors, which have been sampled by drilling. The age of the volcanic bedrock of the oceans is progressively older as one moves

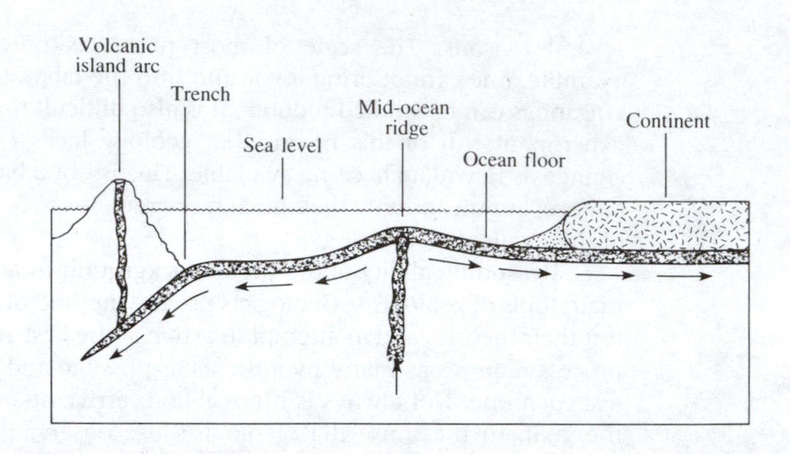

FIGURE I.1
The creation of new sea floor at mid-ocean ridges causes movement of the sea floors. The sea floors return to the interior at convergent plate boundaries, and their melting produces volcanoes. Continents may be carried by the moving sea floors.

away from the ridges in both directions. In the same way, as one travels away from the mid-ocean ridges, the thickness of the sediments above the bedrock is greater and the deepest sediments are older. The conclusion seems inescapable that new ocean floors are being created at the mid-ocean ridges and that the ocean floors are moving away from the ridges. The rate of movement is only a few centimeters per year, but given the earth's great age, the total movement is very large.

If new ocean floor is created at the mid-ocean ridges, then the ocean floors must be consumed somewhere; otherwise, the earth would have to expand (which it does not). The **volcanic arcs** are places where the ocean floor apparently returns to the interior of the earth (Fig. I.1). At many places the volcanic arcs form volcanic island arcs—island chains formed by volcanoes that appear as curves or arcs on a map. Volcanic arcs are geologically active areas with deep earthquakes as well as volcanic activity.

It is believed that the ocean floor returns to the interior of the earth at the volcanic arcs because the earthquakes occur on a plane that slopes beneath the volcanic arc. Movement of a slab of ocean floor could cause the earthquakes. The volcanoes could be the result of partial melting of the slab of ocean floor. Downward movement of the ocean floor is also suggested by the deep submarine trenches in front of volcanic arcs (Figs. I.1 and 3.22). Thus, new ocean floor is created at the mid-ocean ridges and moves slowly toward the volcanic arcs, where it is consumed.

Continental drift and moving sea floors lead to the idea of **plate tectonics.** In plate tectonics, the surface of the earth is believed to be composed of a number of thin slabs or plates that move across the surface (Fig. 3.29). The plates are formed at the mid-ocean ridges, and they return to the earth's interior at **convergent plate boundaries** such as the volcanic island arcs. Where the plates rub each other, earthquakes are also caused by the motion, and these plate boundaries are called **transform faults.**

From time to time these plates collide as they move across the earth's surface. The type of **plate collision** is determined by the composition of the plates at the point of contact. If ocean floor collides with continent, the ocean floor is consumed at the resulting convergent plate boundary. If ocean floor meets ocean floor, one plate is consumed and a volcanic island arc forms. If continent collides with continent, a mountain range is formed (see Fig. 3.55 for illustrations of the three collision types). All of these collisions deform the rocks. The rocks, especially the sedimentary rocks that form in layers, are bent and folded and at places are broken and faulted as well. Igneous rocks are also formed where plates are consumed, and the rising igneous rocks cause metamorphism and deformation. This, then, is the origin of the folded and faulted rocks that form the earth's mountain ranges.

We begin our study by considering the crust of the earth, the outermost few kilometers. First we study the minerals and rocks of the crust, because they tell us much about the composition of the crust and the processes that take place within the crust as well as on its surface. Then we consider erosion—the dominant surface process. Next, we consider the interactions of the processes in the crust and in the deeper parts of the earth and learn how our planet's present surface features formed. This leads us to plate tectonics, a model of the entire earth. Finally, we examine the methods of dating and deciphering the earth's history, and briefly describe the physical history of North America and the development of life as an example of geologic history.

KEY TERMS

Absolute time
Mid-ocean ridge
Volcanic arc
Plate tectonics
Convergent plate boundary
Transform fault
Plate collision

SUPPLEMENTARY READING

Booth, Basil, and Frank Fitch. *Earthshock.* New York: Walker & Co., 1979, 256 pp.

Geikie, Archibald. *The Founders of Geology.* New York: Dover Publications, (1905) 1962, 486 pp. (paperback). Originally published in 1905 by Macmillan & Co.

Howard, A. D., and Irwin Remson. *Geology in Environmental Planning.* New York: McGraw-Hill Book Co., 1978, 478 pp.

Murray, R. C., and J. C. F. Tedrow. *Forensic Geology: Earth Sciences and Criminal Investigations.* New Brunswick, NJ: Rutgers Univ. Press, 1975, 217 pp.

Rabbitt, M. C. *Minerals, Lands, and Geology for the Common Defense and General Welfare, Vol. 1, Before 1879.* (USGS) Washington: U.S. Government Printing Office, 1979, 332 pp. (paperback).

Short, N. M., and others. *Mission to Earth: Landsat Views the World.* NASA SP–360. Washington: U.S. Government Printing Office, 1977, 459 pp.

U.S. Geological Survey. *Earth Science in the Public Service.* Professional Paper 921. Washington: U.S. Government Printing Office, 1974, 73 pp.

1

Composition of the Crust

ELEMENTS—ABUNDANCE DATA

The composition of the crust is shown in Table 1.1. Although there are 80 stable elements, more than 99 percent of the crust is composed of only eight elements.

Before discussing the important implications of these data, we should look at their origin. All abundance data are calculated from estimates of the amounts of the different rock types and their chemical analyses. Obviously, the main problem is estimating the amounts of the rock types. Fortunately, geologic maps showing the distribution of rock types have been made for many parts of the earth's surface.

MINERALS

So far, we have discussed **elements,** *which are the smallest units in nature that cannot be subdivided by ordinary chemical methods. The smallest unit of an element is the* **atom.** The atom can be broken down into even smaller particles such as electrons, neutrons, protons, and so forth, but, except for radioactive elements, this breakdown requires the large amounts of energy available in atom-smashing machines such as the cyclotron.

In nature, atoms are combined to form **minerals,** which can be defined as:

1. *Naturally occurring, crystalline,*
2. *Inorganic substances with*
3. *A definite small range in chemical composition and physical properties.*

TABLE 1.1
Main elements in the continental crust.

Element	Symbol	Weight Percent	Atom Percent	Volume Percent
Oxygen	O	46.40	61.84	94.27
Silicon	Si	28.15	21.40	0.88
Aluminum	Al	8.23	6.51	0.47
Iron	Fe	5.63	2.15	0.48
Magnesium	Mg	4.15	3.64	0.58
Calcium	Ca	2.36	1.26	0.68
Potassium	K	2.33	1.26	1.65
Sodium	Na	2.09	1.94	0.99
Totals		99.34	100.00	100.00

All three conditions must be met for a substance to be a mineral. Minerals are **crystalline** substances; that is, they *have an orderly internal structure,* as opposed to such things as glasses, which are supercooled liquids. A **crystal** *is a solid form bounded by smooth planes that give an outward manifestation of the orderly internal structure* (see Fig. 1.6). Note the distinction between crystal and crystalline—although all minerals are crystalline, they do not necessarily occur as geometric crystals.

The minerals group themselves naturally to form **rock,** which can be defined as:

1. *A natural aggregate of one or more minerals,* or
2. *Any essential and appreciable part of the earth's crust.*

Note that this definition does not require all rocks to be composed entirely of minerals. Some noncrystalline material is present in many quickly chilled volcanic rocks.

Because the rocks of the crust are composed preponderantly of eight elements, we would suspect at first glance that these eight elements, which must form most minerals, would occur in many different combinations. However, most rocks are composed of only a few combinations of elements—or minerals. The reason for this happy situation, which means that a beginner needs to learn only a few minerals to understand most rocks, can be seen in the *volume* data in Table 1.1. Rocks are composed of over 90 percent oxygen by volume, because oxygen atoms (ions) are relatively large. Thus, the number of possible combinations is controlled by the ways that oxygen atoms can be arranged to build minerals, especially because relatively few other atoms are readily available to form minerals. Therefore, although some 3000 minerals have been discovered, only about 20 are common, and fewer than 10 form well over 90 percent of all minerals.

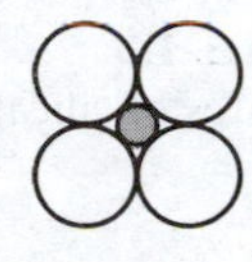

FIGURE 1.1
The relative sizes of atoms determine how they can fit together.

Thus, the key to understanding the structure of minerals lies in understanding how the atoms in a mineral are arranged, or, said another way, how the other atoms are fitted among the oxygen atoms. Clearly, the size of the atoms in question, compared to the size of oxygen, will determine the structure of any mineral. This relationship is shown in two dimensions in Figure 1.1. The number of oxygen atoms that surround an element can be calculated from simple geometric relationships. (Note that in this case the atoms are more correctly called **ions** because *they have gained or lost one or more electrons.*)

As can be predicted from the abundance data, most of the rock-forming minerals are composed largely of oxygen and silicon, generally with aluminum and at least one more of the "big eight." *The sizes of oxygen and silicon are such that the small silicon ion fits inside four oxygen ions.* This unit, which is the building block of the silicate minerals, is called the **SiO_4 tetrahedron** because the four oxygen ions can be considered as situated on the four corners of a tetrahedron with the silicon at the center (Fig. 1.2).

The structures of the silicate minerals are determined by the way these SiO_4 tetrahedra are arranged. Two of these tetrahedra, if joined at a corner, can share a single oxygen ion, forming a very strong bond. With this type of linkage, through sharing of oxygen, the structure of the silicate minerals is formed. The possible structures are shown in Figure 1.3.

Rock-Forming Minerals

We will now look at the more important rock-forming minerals. Their identification will be mentioned here and discussed more thoroughly in the next section. The main rock-forming minerals are all silicates, and their structures are shown in Figure 1.3.

Feldspar **Feldspars** are the most abundant minerals. There are two main types, *plagioclase feldspar* and *potassium feldspar,* and they are complicated by mixing and a number of structural modifications that are too complex to discuss here.

Plagioclase feldspar *is an example of a continuous mineral series.* The high-temperature plagioclase is *calcic,* $CaAl_2Si_2O_8$. At low temperature, *sodic plagioclase,*

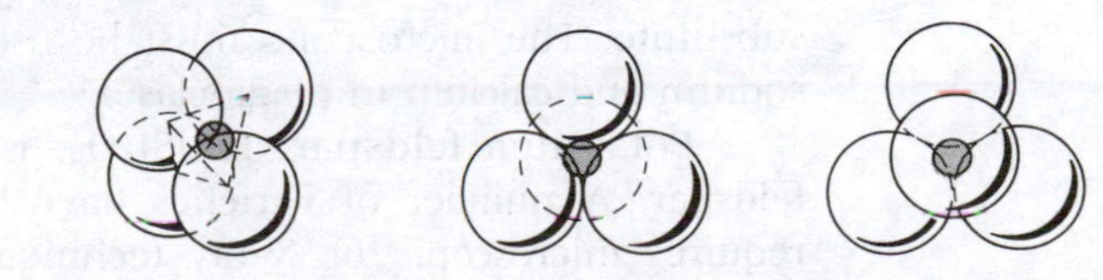

FIGURE 1.2
Three views of an SiO_4 tetrahedron. The large spheres are oxygen, and the small spheres are silicon.

FIGURE 1.3
Structures of silicate minerals.

	Formula of Silicon-Oxygen Unit	*Number of Oxygen Shared*	*Example*
Single tetrahedron	 (SiO_4)	0	Olivine
Double tetrahedron	 (Si_2O_7)	1	Epidote
Ring	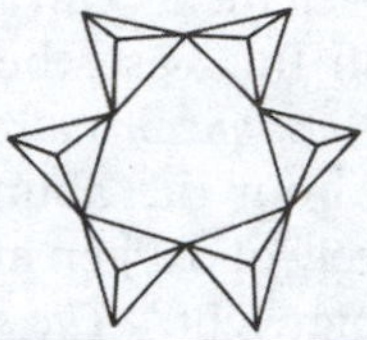 (Si_6O_{18})	2	Tourmaline
Chains	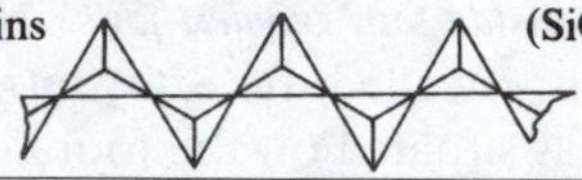 (SiO_3)	2	Pyroxene (augite)
Double chains	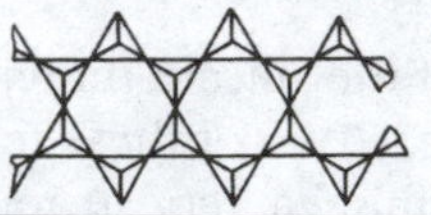 (Si_4O_{11})	2 & 3	Amphibole (hornblende)
Sheets	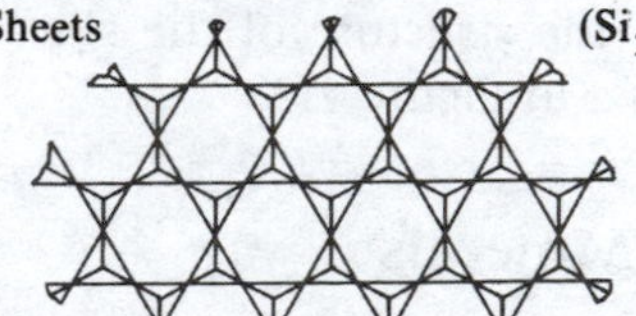 (Si_2O_5)	3	Micas (muscovite, biotite)
Three-dimensional networks	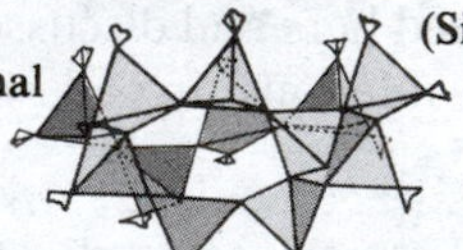 (SiO_2)	4	Quartz, feldspars

$NaAlSi_3O_8$, forms. A continuous series of intermediate plagioclase containing both sodium and calcium can form, depending on the temperature and composition of the melt. This is possible because sodium and calcium ions are about the same size. Aluminum and silicon, which are also nearly the same size, also mutually substitute. The microscope must be used to determine the relative amounts of sodium and calcium in plagioclase.

Potassium feldspar, $KAlSi_3O_8$, is the second, more complicated type of feldspar. A number of varieties have been recognized, but their identification requires microscopic or X-ray techniques. The variety that generally forms in

FIGURE 1.4
Plagioclase striations appear only on one cleavage direction.

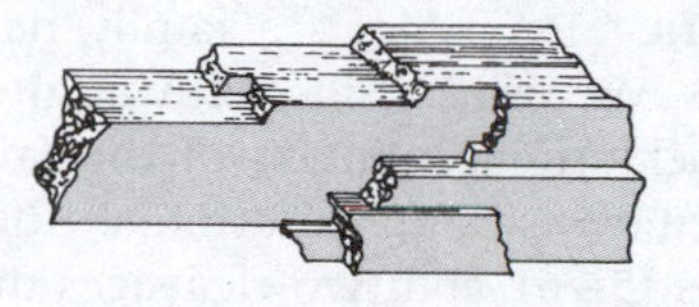

granitic rocks is commonly—although in many cases incorrectly—called orthoclase; the variety in volcanic rocks, which crystallize at higher temperatures than do granitic rocks, is called sanidine.

Limited amounts of sodium can be accommodated in potassium feldspar. This limited mixing is further complicated in that the temperature of formation also controls the amount of mixing.

At high temperatures, such as in volcanic rocks, all mixtures of sodium and potassium are possible and can be preserved by rapid chilling. Slow cooling in the intermediate mixture will allow the potassium feldspar and the sodic plagioclase to unmix. In some rocks the unmixing is complete, and two distinct crystals form; but in many, the *two feldspar crystals are intergrown and form* **perthite.**

Feldspars are identified by their hardness (6), two cleavages at nearly right angles, and light color. Striations on one cleavage direction distinguish plagioclase (Fig. 1.4). Perthite appears as mottled, irregular veining of different colors (Fig. 1.5).

Olivine **Olivine,** $(Mg,Fe)_2SiO_4$, is the simplest of the ferromagnesian or dark-colored minerals. The parentheses in the formula mean that magnesium and iron may be present in varying amounts because their ions are nearly the same size. Magnesium olivine forms at higher temperatures than the iron-rich types. Olivine generally forms small, rounded crystals that may give deceptively low hardness tests because of alteration. Olivine can be recognized by its distinctive olive-green color.

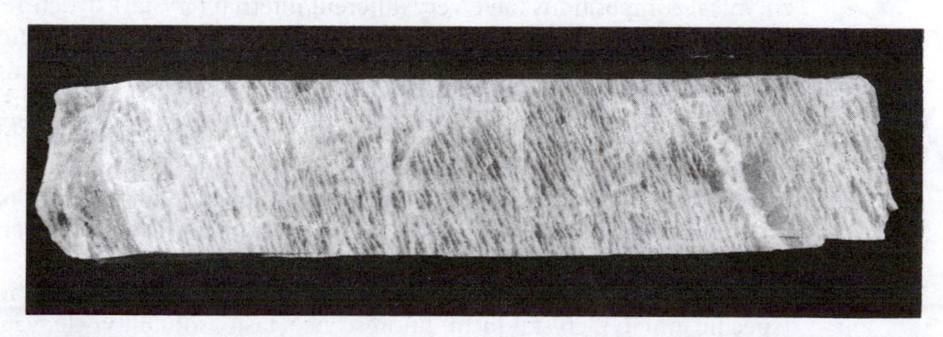

FIGURE 1.5
Cleavage surface of perthite showing the mottled appearance that results from the intergrowth of two types of feldspar. (Photo from Smithsonian Institution.)

Pyroxene Pyroxene is a family name of a very large and complex group of minerals. We will consider only **augite,** $(Ca,Na)(Mg,Fe^{+2},Fe^{+3},Al)(Si, Al)_2O_6$, the main rockforming mineral of the family. The possible range in composition is suggested by its complex formula. Augite is generally recognized by its dark color, hardness (5–6), and two cleavages that meet at nearly right angles.

Amphibole Amphibole is a family name of another very large and complex group of minerals. We will consider only the main rock-former in the group, **hornblende,** $(Ca,Na)_{2-3}(Mg,Fe^{+2},Fe^{+3},Al)_5(Al,Si)_8O_{22}(OH)_2$. Its composition is even more complex than that of augite. Hornblende is black, has a hardness of 5–6, and has two cleavages that meet at about 60 degrees, distinguishing it from augite.

Mica The micas are easily recognized by their colors and perfect cleavage. The common micas and their formulas are:

Biotite—black mica, $K(Mg,Fe)_3(AlSi_3O_{10})(OH)_2$
Muscovite—white mica, $KAl_2(AlSi_3O_{10})(OH)_2$
Chlorite—"green mica," $(Mg,Fe,Al)_6(Al,Si)_4O_{10}(OH)_8$

Quartz Quartz (SiO_2) is generally clear or white but may have any color. It is recognized by its hardness (7) and lack of cleavage. As in many other minerals, the color of quartz is caused by very minor amounts of impurities.

Identification of Minerals

The definition of mineral suggests that minerals can be identified several ways. The fact that minerals have reasonably fixed chemical compositions suggests that ordinary methods of chemical analysis can be used, and historically this approach was important in the development of both mineralogy and chemistry. Two drawbacks to its current use are the need for a laboratory, as well as much skill and knowledge, and the relatively recent discovery that certain minerals with identical chemical compositions have very different internal (crystal) structures. In this latter case, the polymorphs (many forms) generally have very different physical characteristics; for example, graphite and diamond are both composed of pure carbon. Proof that certain minerals were indeed polymorphs had to wait until the use of X-rays to probe crystal structure was perfected. Another problem in the use of chemical analysis is that analysis of the all-important silicate minerals is an extremely difficult task, even in the best-equipped laboratories. Fortunately, this problem can be bypassed by using easily determined physical properties to identify minerals, supplemented where necessary with spot chemical tests.

The physical properties include such things as color, luster, hardness, density (specific gravity), crystal form, fluorescence, taste, solubility, cleavage, magnetism, radioactivity, and many more. The use of the more important physical properties follows.

Color Strictly speaking, **color** *is the color of a fresh, unaltered surface,* although in some cases the tarnished or weathered color may help in identification. For some minerals color is diagnostic, but many, such as quartz, may have almost any color due to slight impurities.

Streak **Streak** *is the color of the powdered mineral.* To see the streak, rub the mineral on a piece of unglazed porcelain called a streak plate. The color of the streak in many minerals is a more constant property than the color of the mineral specimen. The streak may be a very different color from the mineral color, and this fact is of help in identification. Take care, especially when working with small disseminated crystals, that only the mineral and not the matrix is rubbed on the streak plate. Many powdered minerals are white and are said to have no streak. Minerals with a hardness greater than the streak plate will, of course, have no streak.

Luster **Luster** *is the way a mineral reflects light.* The two main types of luster are **metallic** and **nonmetallic.** The distinction is difficult to describe in words; either a mineral looks like metal, or it does not. *A few minerals fall in between and are sometimes called* **submetallic.** Dozens of terms have been proposed to describe all of the types of luster. A simplified outline of the most commonly used terms follows; most are self-explanatory:

Metallic
- Bright
- Dull

(Submetallic)(used only in borderline cases)

Nonmetallic
- Adamantine (brilliant luster, like a diamond)
- Vitreous (glassy)
- Greasy
- Resinous
- Waxy
- Pearly
- Silky
- Dull or earthy

Crystal Form All crystalline substances crystallize in one of six crystal systems (Fig. 1.6). If the mineral grows in unrestricted space, it develops the external shape of its crystal form; if it cannot grow its external shape, its crystalline nature can be determined only under the microscope or by X-ray analysis. Many external forms are possible in each of the systems; however, the system can be determined by the symmetry of the crystal. Crystallography is a fascinating subject, and more information can be found under "Minerals" in the supplementary readings listed at the end of this chapter.

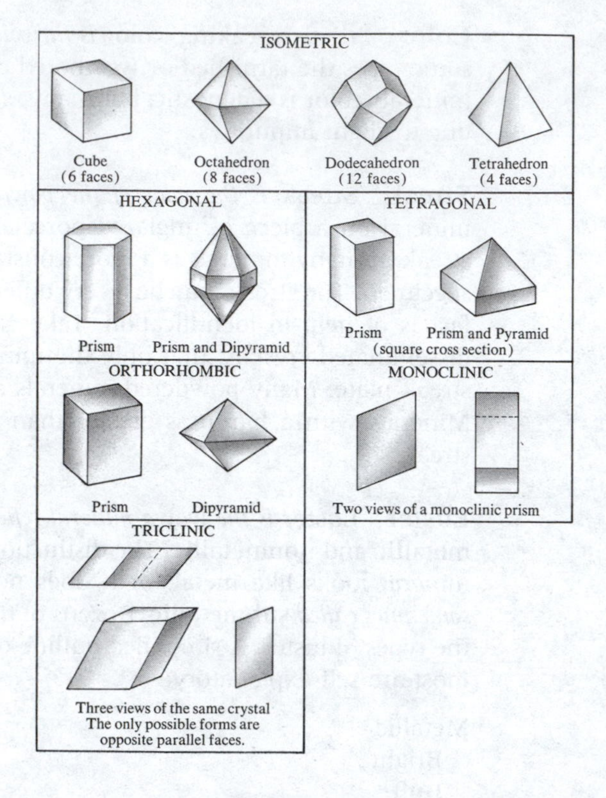

FIGURE 1.6
Examples of the more common crystal forms. They are shown and named for reference only.

Cleavage and Fracture These terms describe the way a mineral breaks. *Any irregular break is termed a* **fracture.** Many terms have been coined to describe fractures, and most are self-explanatory such as even, fibrous, splintery, or hackly. A common type of fracture, **conchoidal,** *is the hollow, rounded type of break that occurs in glass* (see Fig. 1.13B).

A mineral has a **cleavage** *if it has a direction of weakness that, when the mineral is broken, produces a smooth plane that reflects light* (Fig. 1.7). The common minerals may have up to six such directions. Note that a cleavage is a direction; therefore one

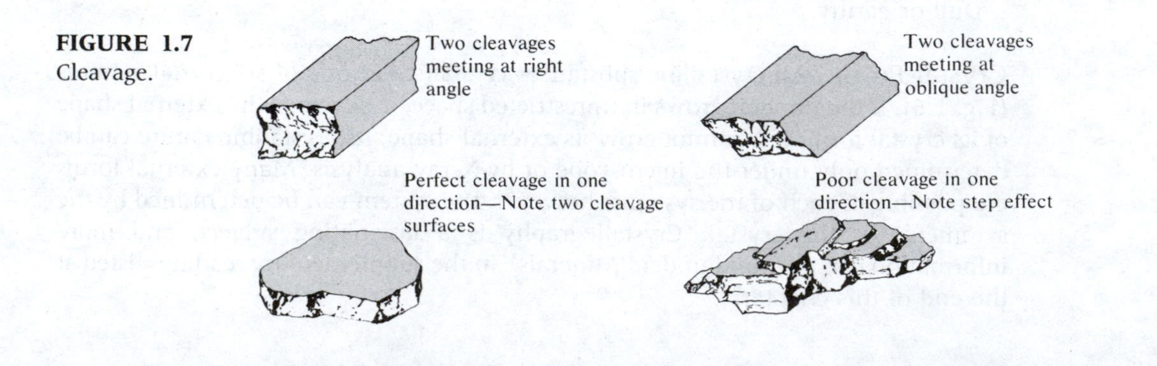

FIGURE 1.7
Cleavage.

cleavage direction will generally produce two cleavage surfaces. Cleavages can be recognized by their smoothness and by their tendency to form pairs or steps, as suggested in the sketches. A series of steps is more common in minerals with poor cleavage—that is, minerals in which the cleavage direction is only slightly weaker than any other direction in the mineral. Even in this case, each of the cleavage surfaces will be smooth and parallel and will reflect light as a unit.

The planes of weakness that are the cleavage directions in a mineral are caused by its atomic structure. Cleavages form along directions of weak bonding, as shown in Figure 1.8.

Hardness The **hardness** *of a mineral is its resistance to scratching.* As might be guessed, it is a difficult property to measure exactly because the amount of force, the shape of the scratches, and the relation of the surface tested to the crystal structure will all affect the measured hardness. In spite of this, the gross hardness is easily measured and of great help in identification. Because surface alteration may change the apparent hardness, only fresh surfaces must be tested.

The hardness scale (Mohs) is:

1. Talc — Softest mineral
2. Gypsum
3. Calcite
4. Fluorite
5. Apatite
6. Orthoclase
7. Quartz
8. Topaz
9. Corundum
10. Diamond — Hardest mineral

(Steps 3–8:) The steps are not of equal value but are arbitrarily defined. The steps are reasonably equally spaced except that the step between 9. corundum and 10. diamond is very large.

A useful hardness scale is:

2.
—Fingernail
3.
—Copper cent (Use a bright, shiny one, or you will only test the hardness of the tarnish)
4.
5.
—Knife blade or glass plate
6. —File (hardened steel)
7. —Quartz
8.

To determine the hardness of a mineral, you must find the softest mineral on the hardness scale that will scratch it. With a little practice, only a knife blade and your fingernail will be necessary to estimate the hardness. Note that two substances of

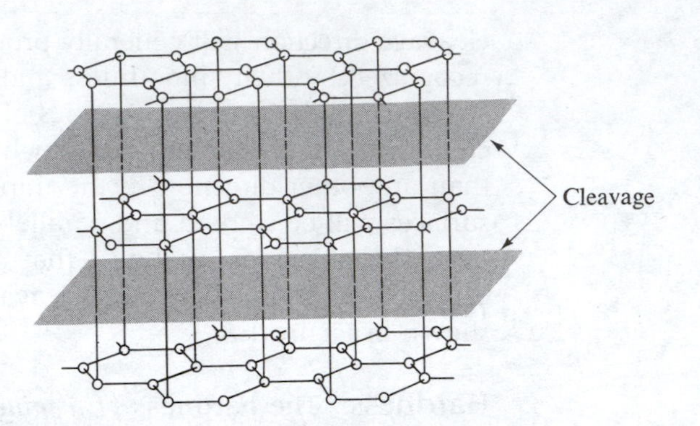

FIGURE 1.8
Structure of graphite. The bonds between the carbon sheets are weak, causing the cleavage in graphite. Graphite and diamond are both composed of carbon only; the differences between them are caused by the internal structure.

the same hardness can scratch each other. A pitfall that can trap the neophyte is accepting any mark on the mineral as a scratch, because just as soft chalk leaves a mark on a hard blackboard, a soft mineral may leave a mark on a hard one. You can check this by trying to remove the mark with your moistened finger, by feeling the mark on your fingernail (even a tiny depression can be detected), and by looking very closely at the mark with a lens. The final test should always be to reverse the test and see whether the mineral can scratch the test mineral.

Specific Gravity **Specific gravity** *is a measure of the relative weight of a substance. It is the ratio of the mass of a substance to the mass of an equal volume of water.* It is measured by weighing the substance in air and in water:

$$\text{specific gravity} = \frac{\text{weight in air}}{\text{weight in air} - \text{weight in water}}$$

This is a difficult measurement to make with most specimens. Estimations of specific gravity can be made fairly accurately by one who is experienced in handling minerals. It is of help in identification, especially with the very dense minerals.

The density of a substance is its mass per unit volume; when expressed in grams per cubic centimeter, density is numerically equal to specific gravity.

Taste and Solubility *Taste* and *solubility* can be determined by touching the specimen with the tongue. A soluble mineral will have the feel of a lump of sugar, and an insoluble mineral may feel like glass.

Reaction with cold dilute hydrochloric acid (HCl) is another solubility test. Calcite will react, producing effervescence. Dolomite (see Tables 1.2 and 1.3) must be powdered to increase the surface area before effervescence occurs. Dolomite will also react with hot dilute acid or with strong acid, both of which also increase the rate of chemical reaction.

Other physical properties may require special equipment to test. Examples are *radioactivity* (Geiger counter), *magnetism* (magnet), and *fluorescence* (ultraviolet light).

Table 1.2 helps to identify the more common minerals. To use the chart, first decide the sort of luster the mineral has (metallic or nonmetallic); second, its color (light or dark); third, its hardness (compared to a knife blade); fourth, whether it has cleavage; and finally, read the brief descriptions to identify it.

ROCKS OF THE CRUST

The rocks of the crust are classified into three types according to their origin. Two of these types are formed by processes deep in the earth and so tell us something about conditions within the crust. They are:

1. **Igneous rocks,** *which solidify from a melt or magma;*
2. **Metamorphic rocks,** *which are rocks that have been changed*—generally by high temperature and pressure within the crust.

The third type, which records the conditions at the surface, is:

3. **Sedimentary rocks,** *which are weathering products of preexisting rocks that are deposited near the earth's surface by wind, water, ice, and biologic activity.*

Much of geology is concerned with the interactions among the forces that produce these three rock types. The relationships are quite involved, but can be illustrated superficially by the **rock cycle** (Fig. 1.9).

Igneous Rocks

Crystallization Igneous rocks are formed by the crystallization of magma. A **magma** *is a natural, hot melt composed of a mutual solution of rock-forming materials (mainly silicates) and some volatiles (mainly steam) that are held in solution by pressure.*

One of the most important concepts in igneous geology is the **reaction series** first described by N. L. Bowen. This series shows the sequence in the crystallization of a basaltic melt, which is the most common type of volcanic rock. It also suggests how *the earliest-formed crystals might be separated from the melt or magma, forming rocks with a composition different from that of the original melt. Such a process is called* **differentiation,** and we will discuss its role when we consider the origin of magmas and of the various types of igneous rocks. Although Bowen originated the idea of a reaction series from studies of experimental melts, the general sequence of crystallization of most rocks was already known as a result of intensive microscopic studies. The reaction series is an oversimplification but is very useful (see Fig. 1.10).

The reaction series really consists of three reaction series of two different kinds. The right-hand side is a **continuous series**; that is, *all compositions of plagioclase from entirely calcic to entirely sodic and all compositions in between occur.* The left-hand side is a **discontinuous series**; that is, *the changes from one mineral to the next occur in discrete steps.* Each of the minerals named in the left-hand series is

TABLE 1.2
Mineral identification key.

Luster	Hardness	Cleavage	Description	Mineral
Metallic luster			Black; strongly magnetic; hardness, 6.	MAGNETITE
			Lead-pencil black, smudges fingers; hardness, 1; one cleavage that is apparent only in large crystals.	GRAPHITE
			Brass yellow; black streak; cubic crystals, commonly with striations; hardness, 6–6.5	PYRITE
			Brass yellow; may be tarnished; black streak; hardness, 3.5–4; massive.	CHALCOPYRITE
			Shiny gray; black streak; very heavy; cubic cleavage; hardness, 2.5.	GALENA
Light-colored nonmetallic luster	Hard—not scratched by knife	Shows cleavage	White or flesh-colored; two cleavage planes at nearly right angles; hardness, 6. Large crystals which show irregular veining are PERTHITE.	ORTHOCLASE (POTASSIUM FELDSPAR)
			White or green-gray; two cleavage planes at nearly right angles; hardness, 6; striations on one cleavage.	PLAGIOCLASE
		No cleavage	White, clear, or any color; glassy luster; transparent to translucent; hexagonal (6-sided) crystals; hardness, 7; conchoidal fracture.	QUARTZ
			Various shades of green and yellow; glassy luster; granular masses and crystals in rocks; hardness, 6.5–7 (apparent hardness may be much less).	OLIVINE
			Any color or variegated; glassy luster; hardness, 5–6; conchoidal fracture.	OPAL
			Any color or variegated; waxy luster; hardness, 7; conchoidal fracture.	CHALCEDONY (AGATE)
	Soft—scratched by knife	Shows cleavage	Colorless to white; salty taste; cubic cleavage; hardness, 2.5.	HALITE
			White, yellow to colorless; rhombohedral cleavage; hardness, 3; effervesces with dilute hydrochloric acid.	CALCITE
			Pink, colorless, white, or dark; rhombohedral cleavage; hardness, 3.5–4; effervesces with dilute hydrochloric acid only if powdered.	DOLOMITE
			White to transparent; three unequal cleavages; hardness, 2.	GYPSUM
			Green to white; feels soapy; one cleavage; hardness, 1.	TALC
			Colorless to light yellow or green; transparent in thin sheets which are very elastic; one cleavage; hardness, 2–2.5 (white mica).	MUSCOVITE
			Green to white; fibrous cleavage; may form veins.	ASBESTOS
		No cleavage	Green to white; feels soapy; hardness, 1.	TALC
			White to transparent; hardness, 2.	GYPSUM
			Yellow to greenish; resinous luster; hardness, 1.5–2.5.	SULFUR

TABLE 1.2
continued

			Description	Mineral
Dark-colored nonmetallic luster	Hard—not scratched by knife	Shows cleavage	Black to dark green; cleavage, two planes at nearly right angles; hardness, 5–6.	AUGITE
			Black to dark green; cleavage, two planes at about 60°; hardness, 5–6.	HORNBLENDE
		No cleavage	Red to red-brown; fracture resembles poor cleavage; brittle equidimensional crystals; hardness, 6.5–7.5.	GARNET
			Various shades of green and yellow; glassy luster; granular masses and crystals in rocks; hardness, 6.5–7 (apparent hardness may be much less).	OLIVINE
			White, clear, or any color; glassy luster; transparent to translucent; hexagonal (6-sided) crystals; hardness, 7; conchoidal fracture.	QUARTZ
			Any color or variegated; glassy luster; hardness, 5–6; conchoidal fracture.	OPAL
			Any color or variegated; waxy luster; hardness, 7; conchoidal fracture.	CHALCEDONY (AGATE)
			Red to brown; red streak; earthy appearance; hardness, 5.5–6.5 (apparent hardness may be less.)	HEMATITE
			Yellow-brown to dark brown, may be almost black; streak, yellow-brown; earthy; hardness, 5–5.5 (may have lower apparent hardness).	LIMONITE
	Soft—scratched by knife	Shows cleavage	Brown to black; cleavage, one direction; hardness, 2.5–3 (black mica).	BIOTITE
			Various shades of green; cleavage, one direction; hardness, 2–2.5 ("green mica").	CHLORITE
			Yellow-brown, dark brown, or black; streak, white to pale yellow; resinous luster; cleavage, six directions; hardness, 3.5–4.	SPHALERITE
		No cleavage	Red to brown; red streak; earthy appearance; hardness, 5.5–6.5 (apparent hardness may be less).	HEMATITE
			Scarlet to red-brown; scarlet streak; hardness, 2–2.5; high specific gravity.	CINNABAR
			Lead-pencil black, smudges fingers; hardness, 1; one cleavage that is apparent only in large crystals.	GRAPHITE
			Yellow-brown to dark brown, may be almost black; streak, yellow-brown; earthy; hardness, 5–5.5 (may have lower apparent hardness).	LIMONITE
			Dark to light green; greasy or waxy luster; some varieties are fibrous; hardness 2–5, generally 4.	SERPENTINE

TABLE 1.3
Characteristics of common minerals.

Mineral	Composition	Color	Luster	Streak	Hardness	Cleavage	Other Properties—Uses
Amphibole—*see* Hornblende							
Asbestos	Hydrous magnesium silicate	Shades of green	Silky	None	Low apparent hardness	Fibrous	Fireproofing and insulation against heat and electricity
Augite	$(Ca,Na)(Mg,Fe^{+2},Fe^{+3},Al)(Si,Al)_2O_6$ Calcium, sodium, magnesium, iron, aluminum silicate	Black–dark green	Vitreous	None	5–6	2 at nearly right angles	Rock-forming mineral
Biotite	$K(Mg,Fe)_3(AlSi_3O_{10})(OH)_2$ Hydrous potassium, magnesium, iron, aluminum silicate	Black–dark green	Vitreous	None	2.5–3	1 perfect	Rock-forming mineral
Calcite	$CaCO_3$ Calcium carbonate	Generally white or colorless	Vitreous to earthy	None	2.5–3	Rhombohedral, 3 not at right angles	Effervesces in cold dilute hydrochloric acid Manufacture of cement
Chalcedony	SiO_2 Silicon dioxide	Any	Waxy	None	7	None	Conchoidal fracture
Chalcopyrite	$CuFeS_2$ Copper-iron sulfide	Brass-yellow—may be tarnished to bronze or iridescent	Metallic—dull	Greenish-black	3.5–4	None	Brittle
Chlorite	$(Mg,Fe,Al)_6(AlSi)_4O_{10}(OH)_8$ Hydrous magnesium, iron, aluminum silicate	Green	Vitreous	None	2–2.5	1 perfect	Rock-forming mineral

TABLE 1.3
continued

Mineral	Composition	Color	Luster	Streak	Hardness	Cleavage	Other Properties —Uses
Cinnabar	HgS Mercury sulfide	Vermillion-red to brownish-red	Adamantine to dull earthy	Scarlet	2.5	1	High specific gravity Only important ore of mercury
Clay	A family of hydrous aluminum silicates which may contain potassium, sodium, iron, magnesium, etc.	Generally light	Earthy	None	Very low apparent hardness	None	Rock-forming mineral
Dolomite	$CaMg(CO_3)_2$ Calcium-magnesium carbonate	Generally some shade of pink, flesh color, or white	Vitreous to pearly	None	3.5–4	Rhombohedral, 3 not at right angles	Powder effervesces in cold dilute hydrochloric acid
Feldspar—*see* Plagioclase *and* Orthoclase (Potassium feldspar)							
Galena	PbS Lead sulfide	Lead-gray	Bright metallic	Lead-gray	2.5	Cubic, 3 at right angles	High specific gravity Ore of lead
Garnet	A family of silicate minerals containing aluminum, iron, magnesium, calcium, etc.	Commonly red	Vitreous	None	6.5–7.5	None	Characteristic isometric crystals
Graphite	C Carbon	Black to steel-gray	Metallic or dull earthy	Black	1–2	1	Greasy feel—marks paper
Gypsum	$CaSO_4 \cdot 2H_2O$ Hydrous calcium sulfate	Colorless, white, gray	Vitreous, pearly, silky to earthy	None	2	1 perfect. Others less good	Plaster of Paris
Halite	NaCl Sodium chloride	Colorless to white	Vitreous	None	2.5	Cubic, 3 at right angles	Salty taste

TABLE 1.3
continued

Mineral	Composition	Color	Luster	Streak	Hardness	Cleavage	Other Properties—Uses
Hematite	Fe_2O_3 Iron oxide	Reddish-brown to black or gray	Earthy to bright metallic	Light to dark Indian-red	5.5–6.5	None	Most important ore of iron
Hornblende	$(Ca,Na)_{2-3}(Mg,Fe^{+2},Fe^{+3},Al)_5(Al,Si)_8O_{22}(OH)_2$ Hydrous calcium, sodium, magnesium, iron, aluminum silicate	Black–dark green	Vitreous	None	5–6	2 at 60°	Rock-forming mineral
Limonite	$FeO(OH) \cdot nH_2O$ Hydrous iron oxide	Yellow-brown to black	Generally earthy	Yellowish-brown	5–5.5	None	Low apparent hardness
Magnetite	Fe_3O_4 Iron oxide	Black	Metallic to submetallic	Black	6	None	Strongly magnetic. Important iron ore
Mica—*see* Biotite, Muscovite, *and* Chlorite							
Muscovite	$KAl_2(AlSi_3O_{10})(OH)_2$ Hydrous potassium-aluminum silicate	Clear–light green	Vitreous	None	2–2.5	1 perfect	Rock-forming mineral
Olivine	$(Mg,Fe)_2SiO_4$ Magnesium-iron silicate	Green	Vitreous	None	6.5–7	None	Rock-forming mineral
Opal	$SiO_2 \cdot nH_2O$ Hydrous silicon dioxide	Any	Vitreous or resinous	None	5–6	None	Conchoidal fracture
Orthoclase (Potassium feldspar)	$KAlSi_3O_8$ Potassium-aluminum silicate	White to pink	Vitreous	None	6	2 at right angles	Rock-forming mineral

TABLE 1.3
continued

Mineral	Composition	Color	Luster	Streak	Hardness	Cleavage	Other Properties —Uses
Plagioclase	$CaAl_2Si_2O_8$ and $NaAlSi_3O_8$ Calcium and sodium aluminum silicate	White to green	Vitreous	None	6	2 at right angles—striations on one cleavage direction	Rock-forming mineral
Pyrite	FeS_2 Iron sulfide	Pale brass-yellow	Metallic—bright	Greenish or brown-black	6–6.5	None	Brittle. Cubic crystals
Pyroxene—*see* Augite							
Quartz	SiO_2 Silicon dioxide	Any	Vitreous	None	7	None	Rock-forming mineral
Serpentine	$Mg_6(Si_4O_{10})(OH)_8$ Hydrous magnesium silicate	Green, may be mottled	Greasy or waxy	None	2–5 4 common	None	Rock-forming mineral
Sphalerite	ZnS Zinc sulfide	Commonly yellow-brown to black	Resinous to adamantine	White to yellow and brown	3.5–4	6	Zinc ore
Sulfur	S Sulfur	Yellow	Resinous	None	1.5–2.5	None	Fracture conchoidal to uneven. Brittle
Talc	$Mg_3(Si_4O_{10})(OH)_2$ Hydrous magnesium silicate	Apple-green, gray, white, or silver-white	Pearly to greasy	None	1	1	Cosmetics

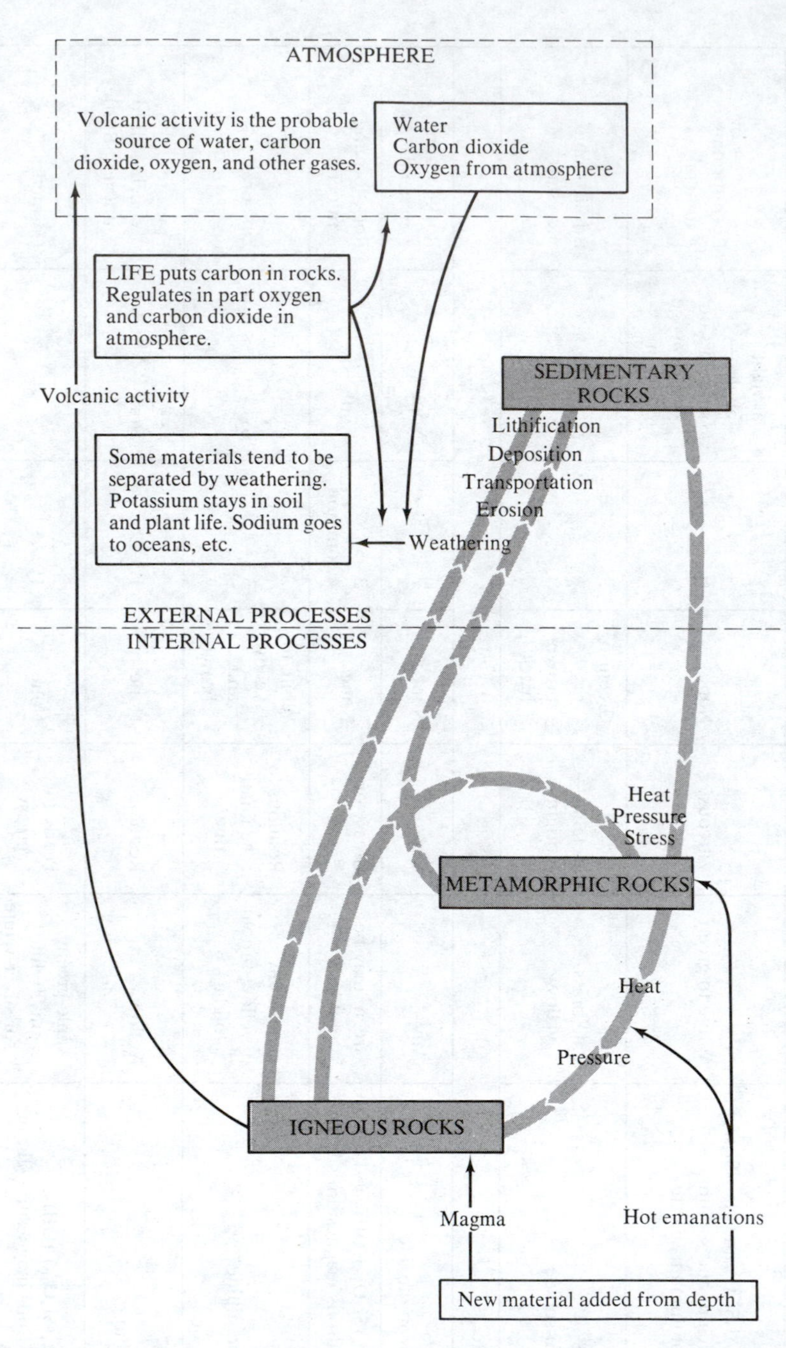

FIGURE 1.9
The middle (shaded part) of this diagram, showing the relationships among igneous, sedimentary, and metamorphic rocks, is what is generally considered the *rock cycle.* The upper and lower parts of the figure show how material is added to and subtracted from the rock cycle. The external processes of the upper half of the diagram are discussed in Chapter 2 of this book, and the internal processes in the lower half are discussed in Chapter 3.

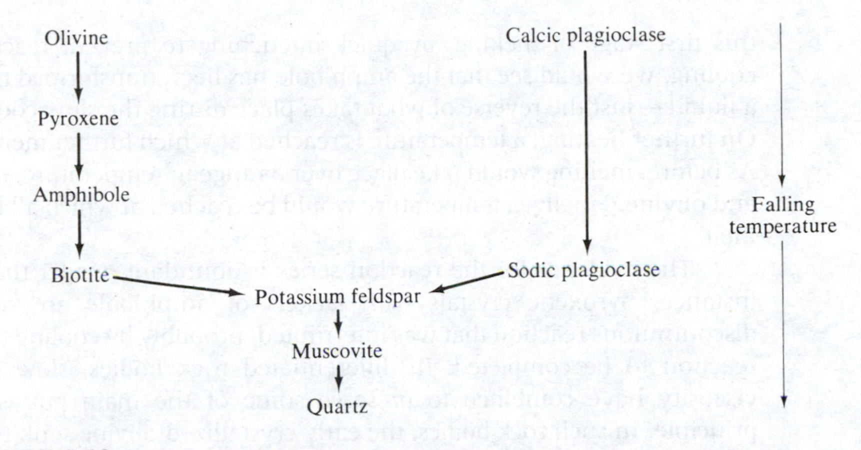

FIGURE 1.10
Bowen's reaction series. Shows the general sequence of mineral formation with falling temperature in a basaltic melt.

actually a continuous series in composition of that mineral group. The lower series is only a sequence of crystallization. The arrangement shows the relative, but not exact, sequence of crystallization; calcic plagioclase and olivine tend to crystallize at the same time, and the lower series is last to crystallize. *It should be emphasized that Bowen's series is only for some basaltic magmas*, but as we shall see basalt magma is very important in igneous geology.

The interpretation of the crystallization of a basaltic magma in terms of Bowen's series is as follows. In general, the first mineral to crystallize from the melt is olivine. In most cases, at the same time or a little later, calcic plagioclase begins to crystallize. The discontinuous series continues with falling temperature; the olivine reacts with the melt to form pyroxene, which at the same time also starts to crystallize directly from the melt. Plagioclase continues to crystallize during falling temperature, but the composition continuously changes to more sodic. If cooling is slow enough, these two processes go on: the minerals on the left side react with the melt to form the next lower mineral while that mineral is crystallizing from the melt, and the plagioclase continuously reacts with the melt to form more plagioclase that is more sodic in composition. These processes end when all the ferromagnesian minerals and plagioclase are formed. Then the potassium feldspar-muscovite-quartz series begins. (This series may in part overlap the sodic part of the plagioclase series.)

Another way to explain the discontinuous series is to consider what happens on heating one of the minerals. When one of the discontinuous series minerals is heated, it does not melt at a single temperature the way ice or similar substances do. Instead, when an amphibole (for example) is heated, it begins to melt when a certain temperature (depending on its composition) is reached, and melting continues over a range of temperature. If we examined the remaining solid during this first stage of melting, by quick quenching to prevent reaction during slow cooling, we would see that the amphibole has been transformed to a pyroxene and a liquid—just the reverse of what takes place during the slow cooling of a magma. On further heating, a temperature is reached at which further melting would occur. As before, melting would take place over a range in temperature, resulting in a melt and olivine. Finally, a temperature would be reached at which all the olivine would melt.

The evidence for the reaction series is abundant. Under the microscope, for instance, pyroxene crystals with jackets of amphibole are seen, recording a discontinuous reaction that was interrupted, probably by cooling too rapidly for the reaction to be completed. In differentiated rock bodies, slow cooling and low viscosity have combined to preserve some of the main points of the reaction principle. In such rock bodies, the early crystallized olivine sank to the bottom and so was removed from contact with the main mass of the magma, allowing progressively lighter-colored rocks to form in the upper parts of the body.

Some believe that such differentiation explains the great variety of igneous rocks, but this is probably an oversimplification. The reaction series applies to basaltic magmas, and the vast areas of basalt lava certainly indicate that basalt is common. Examples of basalt lavas include the Columbia Basin of eastern Washington and Oregon (where at least a thousand meters—3000 feet—of basalt cover tens of thousands of square kilometers), much of Iceland, the Deccan Plateau of India, and the basalt that underlies the ocean. However, rocks of granitic composition are also very abundant, and the continents are underlain by granitic rocks. The vast amount of granitic rocks and their volcanic equivalents suggests that these rocks also represent important magma types because they are far too abundant to be derived from basalt by differentiation. Complete differentiation of basalt can produce, at the most, 15 percent granitic-composition rocks. Understanding the crystallization of basalt gives some insight into the genesis of magmas, as shown in the next section.

Classification *Igneous rocks are classified on the bases of composition and texture.* The classification is an attempt to show the relationships among the various rock types; hence it is an attempt at a natural or genetic classification.

Composition, in the classification of igneous rocks, follows the reaction series. Most rock names came into use long before Bowen described the reaction series. These names had been applied to the more common rocks. It is of more than passing interest that these common groups of rocks fall into the various steps in the differentiation of a basaltic magma according to Bowen's reaction series. This correlation suggests that the compositional part of igneous rock terminology may be genetic, as shown in Figure 1.11.

Because almost all coarse-grained rocks contain feldspar, which is the most common mineral family, the *type* of feldspar is the most important factor in the composition. Of secondary importance is the type of dark mineral which generally accompanies each type of feldspar (see classification diagram, Fig. 1.19).

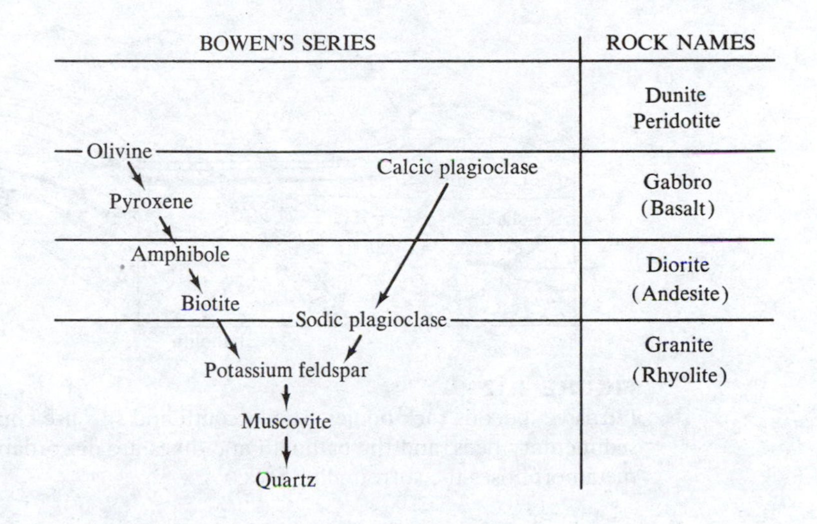

FIGURE 1.11
This diagram shows in a very general way how differentiation by separation of successively formed crystals could result in several rock types originating from a single basalt magma. (Volcanic rock types are in parentheses.)

In a similar way, the **texture** *tells much about the cooling history of an igneous rock.* In igneous rocks the texture refers mainly to the grain size. Rocks that cool slowly are able to grow large crystals; but quickly chilled rocks, such as volcanic rocks, are fine grained. To see the significance, we must consider the mode of occurrence of igneous rocks.

Texture and Mode of Occurrence Igneous rocks occur in two ways, either as **intrusive** *(below the surface) bodies* or as **extrusive** *(surface) rocks.* The ultimate source of igneous magma is probably deep in the crust or in the upper part of the mantle, and the terms intrusive and extrusive refer to the final resting place. Intrusive igneous rocks can be seen only where erosion has uncovered them. They are described as **concordant** *if the contacts of the intrusive body are more or less parallel to the bedding of the intruded rocks* and **discordant** *if the intrusive body cuts across the older rocks.*

Large discordant bodies are called **batholiths** (Fig. 1.12). These are very large features, the size of mountain ranges such as the Sierra Nevada in California, which is nearly 640 km (400 miles) long by about 80 km (50 miles) wide. Many other ranges, both larger and smaller, also contain batholiths. Because batholiths are large, and because they probably were emplaced at least a thousand meters (3000 feet) below the surface, they cooled very slowly. This slow cooling permitted large mineral grains to form; therefore it is not surprising that batholiths are composed mainly of granitic rocks with crystals large enough to be easily seen. As might be expected, batholiths are surrounded by metamorphic rocks. The heat from the crystallizing magma is enough to cause this metamorphism.

Dikes *are tabular, discordant intrusive bodies* (Fig. 1.12). They range in thickness from a few centimeters to several thousand meters but generally are a meter to a dozen or more meters thick. They are generally much longer than they are wide, and many have been traced for kilometers. Most dikes occupy cracks and

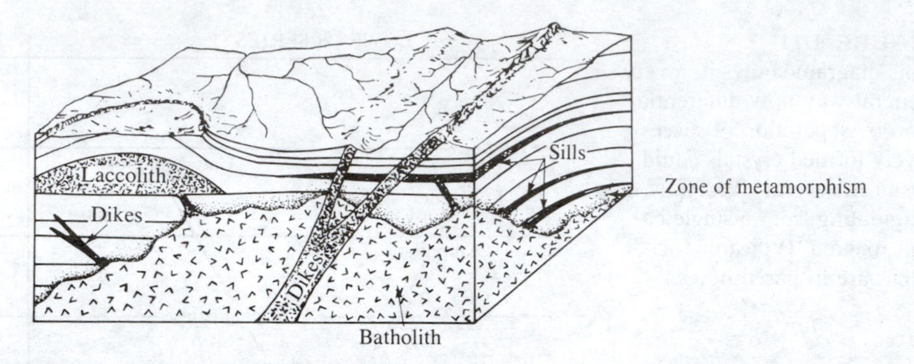

FIGURE 1.12
Intrusive igneous rock bodies. The laccolith and sills are concordant with the enclosing sedimentary beds, and the batholith and dikes are discordant. The heat from the batholith metamorphoses the surrounding rocks.

have straight, parallel walls. Because the intrusive rock is commonly more resistant, dikes generally form ridges when exposed by erosion.

The *concordant intrusive bodies are* **sills** and **laccoliths** (Fig. 1.12). They are very similar and are intruded between sedimentary beds. Sills are thin and do not noticeably deform the sedimentary beds. Laccoliths are thicker bodies that up-arch the overlying sediments, in some cases forming mountains.

Dikes and sills are small bodies compared to batholiths and have much more surface area for their volume; thus, these bodies cool much more rapidly and are commonly fine grained. Laccoliths are generally fine grained but, depending on their size, may be coarse grained or intermediate.

The extrusive rocks are lava flows and other types produced by volcanic activity. These rocks commonly cool rapidly and are fine grained; if cooled so rapidly that no crystallization occurs, they are glassy. Other types of extrusive rocks will be discussed below under the heading "Volcanoes."

Many igneous rocks are *a mixture of coarse and fine crystals,* and this texture is called **porphyritic** (see Fig. 1.13C and D). Such a texture generally records a two-step history of the rock. The large crystals form as a result of slow cooling, perhaps in a deep magma chamber. Then the magma is moved higher in the crust to form a dike, sill, or flow. The remainder of the rock cools rapidly, resulting in a fine-grained or even glassy matrix surrounding the early-formed larger crystals. "Porphyritic" is also used to describe a rock with a coarse matrix and some much larger crystals, but it is doubtful whether such rocks have a two-step history as outlined here.

The temperature of basalt lava is in the range from 1000°C to 1200°C (1800°–2200°F), felsite lava is about 900°C (1600°F), and granitic rocks are believed to be emplaced at about 820°C (1500°F). The cooling time of igneous rocks is not well known. Batholiths are believed to solidify in one million to a few million

FIGURE 1.13
Examples of some types of igneous, metamorphic, and sedimentary rocks.

A. **Granite.** Interlocking texture formed by crystallization from a melt.
B. **Obsidian** or volcanic glass. The round, shell-like fractures are called conchoidal fractures.
C. **Porphyritic andesite.** The large, dark hornblende crystals grew slowly, and rapid chilling produced the fine-grained groundmass.
D. **Porphyritic texture** with large feldspar crystals.
E. Small, light-colored **dikes** cutting a dark rock.
F. **Sandstone.** Bedding is too thick to show in this specimen.
G. **Conglomerate.** Rounded pebbles in a sandstone matrix.
H. **Breccia.** The angular pebbles in this rock probably were not transported far. Compare with G.
I. **Fossiliferous limestone.** A fine-grained white limestone with abundant fossil fragments.
J. **Marble.** Coarse interlocking texture.
K. **Gneiss.** The foliation is produced by discontinuous layers of dark- and light-colored minerals.
L. **Gneiss.** Similar to K but with more irregular layers.

(Photos courtesy Ward's Natural Science Establishment, Inc., Rochester, NY)

years. Dikes, depending on size, probably solidify in several tens of thousands of years. Basalt lava is known to have hardened to a depth of more than 3 m (10 feet) in two years. Rocks are poor conductors of heat; hence intrusive rocks take a very long time to cool because the heat must be conducted away by the surrounding rocks. Lava flows lose their heat to the atmosphere much more quickly.

A summary of textural terms is useful in learning to recognize igneous rock types:

Glassy— no crystals
Fine-grained—grains can be seen in sunlight with a good hand lens
Coarse-grained—crystals easily seen
Pegmatitic—crystals over one centimeter. (This texture is discussed in the next section.)
Porphyritic—rocks generally named for the texture of the groundmass, giving such terms as *porphyritic basalt* for a dark, fine-grained rock with some larger crystals and *porphyritic granite* for a granite with some much larger crystals.

Volcanoes A great deal can be learned about igneous processes from study of volcanoes. They are, after all, the only direct evidence we have for the existence of magma in the crust. It is possible to discuss here only a few of the more significant features of volcanoes.

The active volcanoes today are of several different types. The differences among them seem to depend on the composition of their magmas, particularly on the volatiles in the magmas. *The main volatile in magma is water,* which escapes from volcanoes in the form of steam. Carbon dioxide is also a common volcanic gas, but the sulfur gases (hydrogen sulfide and the oxides of sulfur) are the gases most easily noted near volcanoes because of their strong odors.

A simple analogy illustrates the behavior of water in volcanic magma. Most magmas are believed to contain at least a few percent water in solution. The water is dissolved in the magma just as carbon dioxide gas is dissolved in a bottle of soda pop. Most magma is able to dissolve this water because the magma is under considerable pressure while deep in the earth, just as gas is dissolved in the soda pop as a result of the pressure inside the bottle. When the bottle is opened, the pressure is reduced and the gas begins to bubble out of the liquid. In the same way, when magma is erupted to the surface, the confining pressure is reduced, and the gas—in this case steam—is released.

If the magma has a low viscosity (is more fluid), the gas merely escapes or forms bubbles in the resulting rock. The tops of lava flows can commonly be distinguished from the bottoms by the presence of these bubbles. In some cases the bubble holes are filled by material deposited by these fluids, or are filled at a later time. The fillings are commonly silica materials such as opal or chalcedony, but many other minerals also occur here.

If a magma is erupted very suddenly it may form a *froth, which quickly solidifies to form the light, glassy rock* **pumice.** As a result of the many bubble holes, pumice is light enough to float. The formation of pumice can be likened to the opening of a bottle of warm pop or champagne, which results in the production of a froth.

More violent results occur when more viscous (less-fluid) magmas are erupted to the surface and the gas attempts to escape. In this case the plastic rock is fragmented by the pressure of the expanding gas. The fragments, or **pyroclasts,** produced in this way are named according to their size and shape (Table 1.4). Pyroclasts smaller than 2 mm are called **ash grains,** those between 2 and 64 mm are termed **lapilli,** and those larger than 64 mm (about 2 1/2 in.) are called **blocks** if they are angular and **bombs** if they are rounded or show other evidence of having been melted or partly melted when they were ejected. (The term *ash* implies burning, but of course these rocks do not result from burning. The term was first applied when volcanoes were thought to result from subterranean burning, and so is a historical mistake.) Unconsolidated deposits of these materials are called **tephra.** The rocks formed by this process are termed **pyroclastic** ("fire-broken") **rocks.**

Pyroclastic rocks are named according to the size of the fragments without reference to composition. Consolidated ash is called **tuff;** consolidated lapilli is called **lapilli tuff;** a pyroclastic rock composed of blocks is called **pyroclastic breccia;** and one compounded of bombs is called **agglomerate.** The main identification problem with these rocks is distinguishing them from normal sedimentary rocks formed by erosion of a volcanic area. The latter rocks are called **epiclastic tuff, epiclastic breccia,** or the like, depending on the size of the fragments. In general, field study is needed to distinguish between these two types of accumulation.

Pyroclastic eruptions can be quite violent, and the expanding bubbles may completely shatter the pumice, producing a mixture of hot gas and glass fragments.

TABLE 1.4
Classification of pyroclastic material.

	Fragment	Unconsolidated Deposits	Rocks
General term	PYROCLAST—fragment ejected by volcanic action	TEPHRA	PYROCLASTIC ROCK
Pyroclasts>64 mm	BLOCK—angular, solid when ejected	BLOCK TEPHRA	PYROCLASTIC BRECCIA
	BOMB—rounded or other evidence suggesting melting or partial melting when ejected	BOMB TEPHRA	AGGLOMERATE
Pyroclasts 2–64 mm	LAPILLUS	LAPILLI TEPHRA	LAPILLI TUFF
Pyroclasts<2 mm	ASH GRAIN	ASH	TUFF

Note: The term *epiclastic* is used if the rock or deposit is composed largely of material eroded from volcanic terranes, that is, epiclastic tuff.

Such a mixture is quite fluid because of the gas and can move very rapidly down even a gentle slope. When it comes to rest, the hot glass fragments weld together, forming a **welded tuff.** Many such rocks closely resemble obsidian or felsite, and have about the same composition as the batholiths. Their wide and voluminous occurrence has been recognized only recently through microscopic and field studies. Welded tuffs were first recognized in 1902 when an eruption from Mount Pelée on Martinique Island in the Caribbean Sea wiped out the city of St. Pierre, killing all but about four of its 25,000 to 30,000 inhabitants. One who was not killed was a prisoner who was saved by the thick walls of his cell.

Other types of volcanic eruption range from quiet to violent. The quietest are the basaltic **shield volcanoes,** *so called because the extremely fluid magma builds very gentle slopes that resemble shields* (Fig. 1.14). The Hawaiian Islands are examples of this type of quiet eruption; their most violent displays are fountains of incandescent lava thrown up from lava lakes. Despite their gentle slope, the Hawaiian Islands rise 4600 m (15,000 feet) from the floor of the Pacific plus another 4000 m (13,000 feet) above sea level.

Most other volcanoes combine explosive eruptions with outpouring of lava to build **composite cones** composed of both lava and pyroclastic rocks (Fig. 1.14). Such volcanoes are high mountains generally composed of andesite, such as Mount Shasta in California and Mount Rainier in Washington State. Many differences in mode of eruption exist among the composite volcanoes. The Italian island of Stromboli has mild explosions that produce showers of bombs and so has built a cinder cone with some lava. At the other extreme of explosive volcanoes are Crater

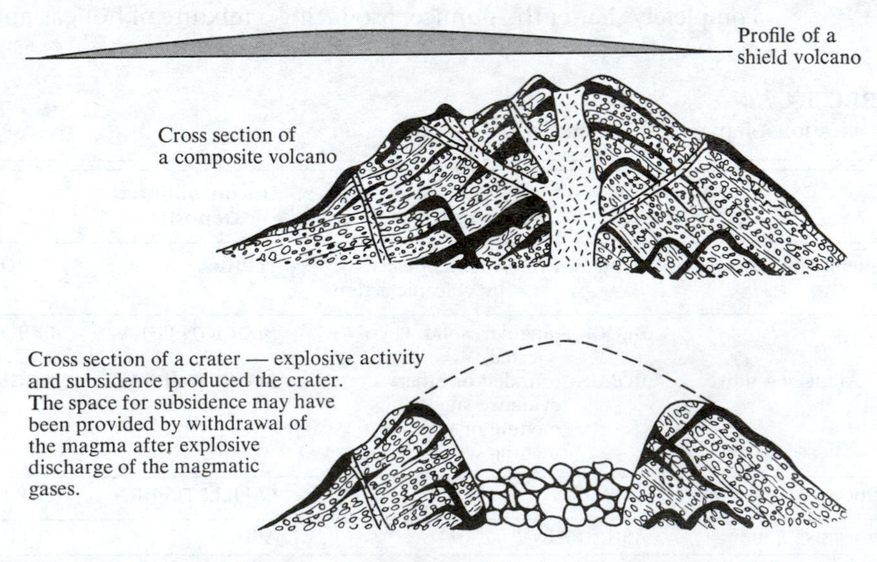

FIGURE 1.14
Types of volcanoes.

Lake, Oregon, and Krakatoa in the East Indies. Both of these volcanoes exploded violently, removing the top of the mountain by a combination of explosion and subsidence. This is the origin of Crater Lake, and the extent of the former mountain can be determined by projecting the slope of the sides of the present mountain (Fig. 1.14). In 1883, the 800-m high (2600 feet) island of Krakatoa, which was part of an earlier, even larger volcano, exploded violently, destroying the island and causing giant waves that caused great damage in nearby Java and Sumatra.

Mount Saint Helens in the Cascade Range in southern Washington State erupted in 1980, causing widespread damage. Its last previous eruption was in 1857, and at that time it had been active since about 1800. Study of Mount Saint Helens and its deposits shows that it has erupted 15 times in the last 4500 years, and that a typical eruptive period lasts about 100 years, with 100–500 quiet years between eruptive intervals.

The eruption in 1980 began with a number of shallow, mild earthquakes. A week later, on 27 March, steam issued from the summit. The earthquakes and eruptions continued, with rock and ice carried at times to heights of more than 5 km (3 miles). Fine rock material was carried many kilometers, and sulfur dioxide as well as steam issued from the crater. During this time, the crater enlarged, and an unusual type of earthquake activity called *harmonic tremors* revealed that magma was moving in the volcano. By late April, the north side of the crater was rising and bulging northward. Apparently, a dome was being emplaced, and the uplift continued at more than a meter (3 feet) per day (Fig. 1.15A).

The big eruption, which occurred on 18 May (Fig. 1.15B), began with two closely spaced earthquakes that caused the bulge, now 125 m (400 feet) high, to landslide. The landslide released the pressure on the dome or plug, and a blast of incandescent gas and ash resulted. The blast was estimated at 50 meters (160 feet) per second, and it devastated an area of 500 km^2 (200 mi^2). It knocked down trees and seared them with 200°C (400°F) temperatures. Between 2 and 3 km^3 (0.75–1 mi^3) of landslide and pyroclastic material moved as much as 20 km (12 miles) down stream valleys, filling some to depths of 60 m (200 feet). After the initial blast, the ash and steam rose vertically to heights up to 19 km (12 miles) and were carried hundreds of kilometers to the east. Fine ash several centimeters deep accumulated at many places downwind. A few hours after the blast, mudflows and debris flows fed by melting glaciers moved into river valleys, causing some flooding and warming of the river water (Fig. 1.15D). The blast removed some 400 m (1300 feet) from the top of Mount Saint Helens, and enlarged the summit crater to a diameter of about 2 km (1.2 miles) and a depth of 1.5 km (1 mile) (Fig. 1.15C).

The water and other volatiles in magma may also be responsible for the formation of veins containing ore minerals. In a deep intrusive body such as a batholith, the rim is quickly chilled and solidifies first. This is commonly recorded by the fine-grained margin of an intrusive body. In this manner the volatiles are trapped within the intrusive body; and as the body solidifies, presumably from the margin inward, the volatiles are confined to a smaller and smaller volume in the center of the body. At the same time, the tiny percentage of minor elements in the

A

B

C

D

FIGURE 1.15
1980 eruption of Mount Saint Helens. A. On 10 April, before the eruption, the mountain was symmetrical and snow-covered. B. Eruption on 18 May. C. Summit area after the eruption. D. Mudflow fills the Toutle River valley after the eruption. (Photos courtesy U.S. Geological Survey; parts A, B, and D by Austin Post)

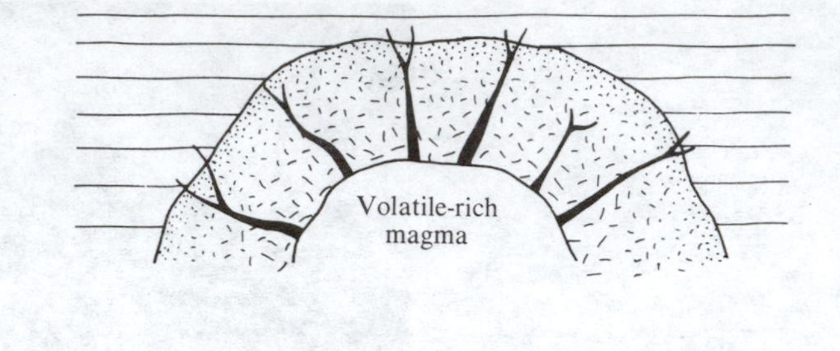

FIGURE 1.16
Cross section of a crystallizing batholith, showing origin of veins. Crystallization has proceeded inward from the fine-grained, quickly chilled margin. The volatiles are concentrated in the still-molten center. The increased pressure causes cracks to form that are filled by the metal-bearing quartz-rich volatiles.

magma which do not fit into the crystal structures of the rock-forming minerals, such as copper, lead, zinc, gold, silver, and sulfur, are also concentrated with the volatiles. The total amount of metallic elements with the volatiles may be quite large, especially in the case of a batholith.

This fluid is also rich in quartz, which might be expected because it is the last mineral to form in the reaction series. This residual fluid eventually escapes, perhaps because of the increase in pressure that can result when the volatiles are concentrated in a small volume. The increase in pressure may cause cracks that are filled by this material, forming the quartz veins that contain mineral deposits (Fig. 1.16). Mineral deposits form in many different ways, and the process outlined here is only one hypothesis.

Another possible effect of the fluids in an intrusive body is to reduce the viscosity of the magma, thus increasing the mobility of the ions in the magma. Acting in this manner and perhaps in other ways, the fluid may enable very large crystals to form. The resulting *very-coarse-grained rocks are called* **pegmatites,** and they may have the same composition as the enclosing magma or may contain unusual minerals. In the latter case the fluids are believed to concentrate the minor elements to form the unusual minerals. Pegmatites commonly occur as dike-like masses in or near intrusive bodies. They are actively sought because they contain large and often beautiful minerals, such as beryl, tourmaline, topaz, fluorite, all of the rock-forming minerals, and many more.

In still other cases these fluids may cause widespread alteration of the intrusive body and/or the surrounding rocks. Much more study is needed to understand the actions of the magmatic volatiles that are believed to cause all of these diverse effects.

Volcanic Hazards The destruction caused by volcanoes is of two kinds—the direct destruction by material issuing from the volcano and the indirect damage caused by mudflows and floods. The former has already been described in the discussion of explosive volcanoes, such as Mount Pelée and Krakatoa. There is little defense against such rapid destruction except evacuation at the first indication of activity by the volcanoes. The first eruption may be of a type called the "glowing cloud" *(nuée ardent)* and may occur anywhere on the volcano. In such cases there may be no escape. In the quieter eruption of basaltic volcanoes, such as those in the Hawaiian Islands, the fluid basalt moves at a few kilometers an hour, and generally there is enough time to evacuate. This may seem of little value if your home and fields are inundated by either lava or airborne ash, but at least you are alive and able to start over!

Indirect damage from volcanoes can also be swift and without warning. Many volcanoes are high mountains and are glacier covered, even in the tropics. An eruption under a glacier can cause extensive melting that results in floods. Also, many volcanoes, especially those that have been inactive for long periods, have large lakes in their depressed summit areas. Crater Lake in Oregon is an example. An eruption that breeches the lake, or an eruption into the lake, can cause the lake to drain and so cause flooding. The eruption of a submarine or shallow-island volcano such as Krakatoa can cause flooding by seismic sea waves.

Probably the largest areas inundated by volcanoes are those swept by mudflows. Although some mudflows are concurrent with the eruption, many occur at the next rainy season. Although the latter mudflows can be predicted, lives are lost because many people refuse to leave their homes and fields. Mudflows are probably the most important way that pyroclastic volcanic rocks are distributed from volcanoes and so are a common geologic process. A pyroclastic eruption can deposit ash and coarser material on the slopes of the volcano and on other slopes downwind. This loose material is not stable and at the next rain becomes mud that can flow even on very gentle slopes. Rain commonly accompanies volcanic eruption, and, although we do not know why, it may be related to the large amounts of steam that are part of most eruptions. Thus, mudflows can accompany eruptions. Pyroclastic eruptions commonly kill vegetation by covering low plants, and the falling debris strips the leaves from trees. Without vegetation, the runoff is increased, and the possibility of mudflows is greater.

Origin of Magma and Plate Tectonics Igneous rocks must originate deep in the earth. To find out how and where magmas form, volcanic areas should be studied. The volcanic regions associated with plate tectonics are active geologically and are a good place to start.

At the mid-ocean ridges, new basaltic ocean floor is created. Submarine volcanism there has been observed and photographed by geologists working in small research submarines. At a few places, the mid-ocean ridge rises above sea level, forming islands such as Iceland where basalt volcanism can be observed. The mid-ocean ridges are also areas of high heat flow. Heat-flow measurements taken all over the earth reveal that, at the mid-ocean ridges, the flow of heat energy from the interior is about twice the average of the rest of the ocean. This localized heat flow apparently is the cause of the mid-ocean ridges and their volcanic activity.

The ocean-floor basalt is about 5 km (3 miles) thick, and this layer and the overlying sedimentary and volcanic rocks are called the **oceanic crust.** The layer below the crust is called the **mantle.** It is from this layer that the oceanic basalt must originate. The mantle is believed to be composed largely of the minerals olivine and pyroxene, which, depending on the amounts of each, form the rock called **peridotite,** or **dunite.** Seismic studies show that in parts of the upper mantle, the rocks are soft because seismic waves are attenuated. The temperature and pressure at that level are such that the mantle rocks are partially melted. At the places where the heat flow from the deep interior is high—the mid-ocean ridges—more of the rocks may be melted. Rocks expand when they are heated and when they are melted, and this expansion may be the reason for the mid-ocean ridges. The hot rocks take up more volume, so the upper mantle pushes the sea floor upward, forming the mid-ocean ridges. The mantle rocks below the ridges, although partially melted, are still quite rigid (Fig. 1.17).

The origin of the basaltic magma that rises to the surface at the mid-ocean ridges is probably the partially melted material from the upper mantle. Rocks do not have a single melting point as simpler materials do—ice and copper, for instance, at any given pressure, will each melt at a certain temperature. Many minerals and most rocks behave differently, and melting occurs over a wide temperature range. Partial melting of the mantle peridotite can form a magma of basaltic composition. A basalt magma formed in this way would be less dense than the rest of the mantle and so would rise to cause submarine volcanism on the ocean floor.

The volcanic island arc volcanoes erupt lava of andesite composition. This magma, too, may be caused by partial melting. At the volcanic island arc–trench areas, seismic studies show that the basaltic ocean crust and its overlying sediments are carried deep into the upper mantle (Fig. 1.18). In this case, the basalt and

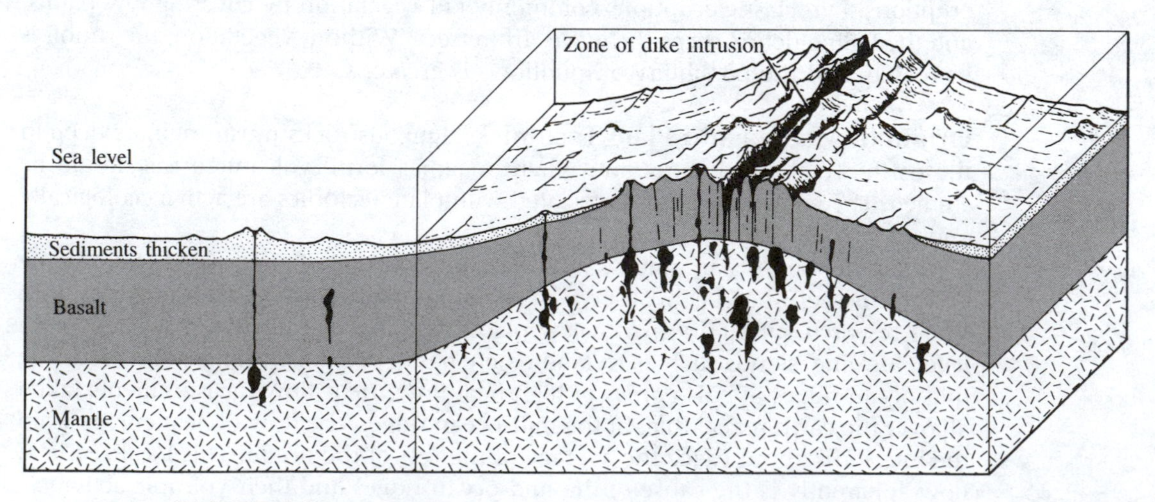

FIGURE 1.17
Ocean-floor basalt originates at mid-ocean ridges by partial melting of mantle rocks.

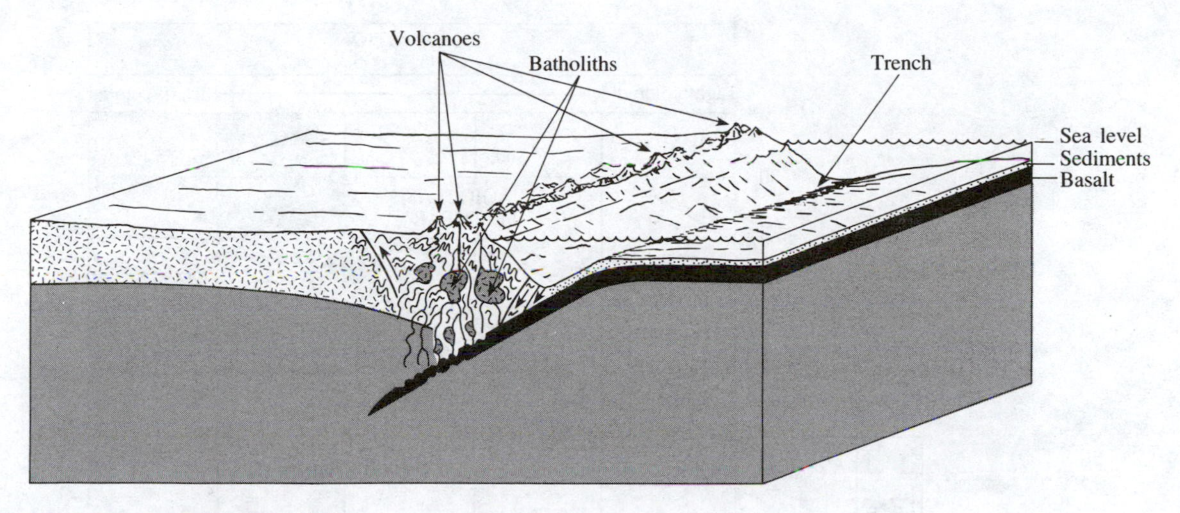

FIGURE 1.18
Batholiths and volcanoes form at a convergent plate boundary from partial melting of ocean-floor basalt and sedimentary rocks.

sediments are partially melted by the higher temperature and pressure of the upper mantle into which they are pushed. The magma generated in this way may not always reach the surface and cause volcanic activity. These less fluid magmas may form intrusive bodies such as batholiths, as suggested in Figure 1.18.

Identification and Interpretation Identification of the coarse-grained igneous rocks presents no serious problems if one is able to identify the minerals and estimate their relative amounts. The first clue comes from the *color* of the rock; the chart (Fig. 1.19) will suggest what minerals might be found. In the use of the term "light color," understand that any shade of red or pink, no matter how dark, is considered a "light" color. In the same way, any shade of green is considered a "dark" color. Overall color cannot be used in place of good mineral determination because exceptions are common enough to be troublesome (especially on examinations).

In a sense, the fine-grained rocks are more difficult to identify than are the coarse-grained rocks because less information in the form of identifiable minerals is available; but because of this, the classification also is less rigid. The procedure is similar in that color is used as the first step and, if no porphyritic crystals are present, the only step. If the porphyritic crystals can be identified, they are used in the same way as are the identifiable minerals in coarse-grained rocks; but because all potential minerals may not be present, color must be considered in deciding on a name. Care must be used not to mistake filled gas holes for porphyritic crystals. In many cases, it is necessary to have a chemical analysis in order to name a fine-grained rock accurately. Also, the volcanic rocks are not simply fine-grained

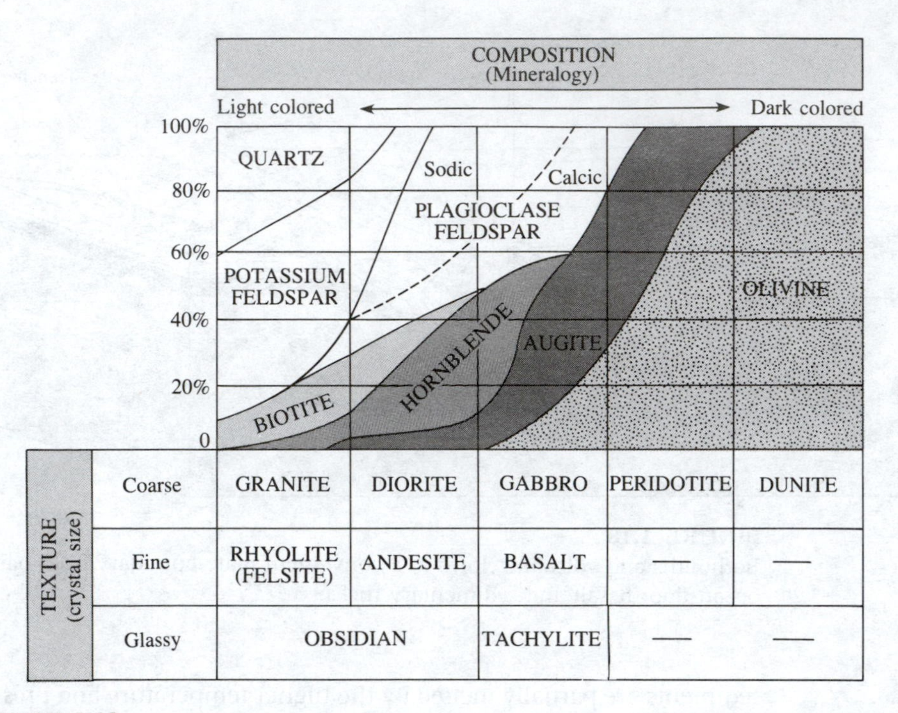

FIGURE 1.19
Chart showing the classification of the igneous rocks. Composition is indicated horizontally, and texture is indicated vertically. The upper part of the figure shows the range in mineral composition of each rock type.

equivalents of the coarse-grained rocks, as the chart might lead one to believe, but contain some small, distinct variations in composition, which chemical analyses will detect.

On the chart shown as Figure 1.19, granite is used for all of the coarse-grained, quartz-bearing rocks. A number of other names are also in use for these rocks, depending on whether they contain one feldspar, two feldspars, or mixed feldspar (perthite).

Understanding the way a rock formed is much more important than simply naming it. Therefore, when studying an igneous rock, try to interpret its *origin*, mainly following the discussion in the section on texture and mode of occurrence. Thus, the coarse texture of a granite implies that the specimen came from a large, slowly cooled intrusive body, but a pumice sample must have come from a volcano, perhaps from an explosive eruption. When actual outcrops are studied, all of the information discussed here can be used to decipher the origin and history of the rocks.

Sedimentary Rocks

Sedimentary rocks form at or near the surface of the earth. Most are deposited in beds or layers. They comprise a very small volume of the earth, only about 5 percent of the crust. In spite of their small volume, however, they cover about 66 percent of the surface. They are formed in a number of different ways, but because the raw materials come from weathering, we will begin there.

Weathering **Mechanical weathering** *breaks up a preexisting rock into smaller fragments, and* **chemical weathering,** *acting on these small fragments, rearranges the elements into new minerals.*

Mechanical weathering is done mainly by the forces produced by the expansion of freezing water and, to a lesser extent, by growing roots, burrowing animals, lightning, and work of humans. The most important of these methods is the result of the nine-percent volume expansion that water undergoes when it freezes. Tremendous forces are produced by this expansion **(frost wedging),** and their efficacy is shown by the thick layers of broken rock that form in areas that go through a daily freeze-thaw cycle many times a year (Fig. 1.20).

The products of mechanical weathering include everything from the huge boulders found beneath cliffs to particles the size of silt. The erosion and transportation of these fragments will be discussed in Chapter 2 on erosion. These fragments can form rocks such as sandstones, and the shape of the fragments in such a rock can reveal some of its history. Thus, if the fragments are sharp and angular, they probably were buried quickly; but if the fragments are rounded, they were abraded during long transportation (Fig. 1.21).

The rate of chemical weathering depends in general on the temperature, the surface area, and the amount of water. Except in cold or very dry climates, chemical weathering generally keeps up with mechanical weathering, and the two can be

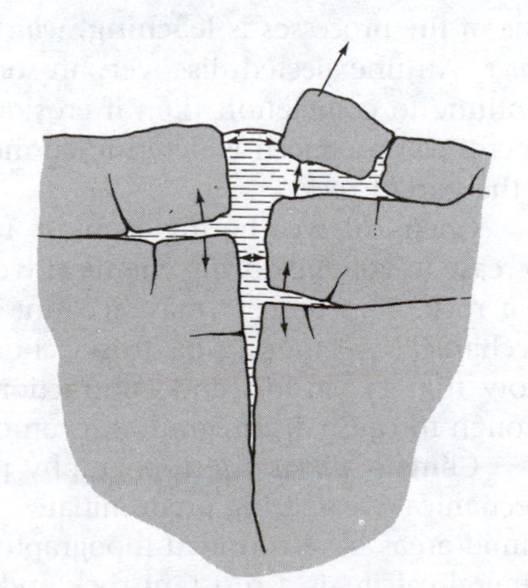

FIGURE 1.20
Frost wedging. Water in crack freezes from top down because ice is less dense than water (floats) and because the water in a crack is cooled most at the surface. This restricts the water in the crack. Because ice occupies 9% more volume than water, further freezing creates forces due to this expansion. These forces cause further cracking of the rock.

FIGURE 1.21
Sand grains showing different degrees of roundness. The rounded sand grains were probably shaped by abrasion during longer transportation than that to which the angular grains were subjected.

separated only in concept. Mechanical weathering provides the large surface area necessary for chemical activity to take place at the surface of the earth (Fig. 1.22).

In following chemical weathering, we will see what happens to the elements in the main rock-forming minerals. The main reactions involved are *oxidation, hydrolysis,* and *carbonation.* **Oxidation** *is reaction with oxygen in the air to form an oxide,* **hydrolysis** *is reaction with water,* and **carbonation** *is reaction with carbon dioxide in the air to form a carbonate.* The carbonation reaction begins with the uniting of carbon dioxide (CO_2) and water (H_2O) to form carbonic acid (H_2CO_3), or $CO_2 + H_2O \rightarrow H_2CO_3$. This acid plays an important role in many weathering reactions. In all of these reactions, substances are added so that the total volume is increased.

The details of how these reactions take place are not well understood and are an important area of current research in soil science. The reactions take place slowly and are probably influenced by the composition of the water films that surround the particles and by the organic materials present (particularly the acids). One of the processes is **leaching,** *which is the removal of soluble materials by ground water.* An unexpected discovery in soil science is that if the reactions are able to continue to completion (i.e., if erosion does not remove the immature soil), the type of soil produced is determined more by the climate than by the composition of the parent rock.

Chemical weathering is most intense on grain surfaces. Because of the increase in volume during chemical weathering, this trait can cause disaggregation of a rock. This process may account for fragmentation of rocks in areas where mechanical weathering by frost action does not occur. Laboratory experiments show that expansion and contraction from daily heating and cooling are not enough to cause disintegration, even on a desert.

Climate affects the topography produced by weathering. Dry areas, where mechanical weathering predominates, generally have bold, angular cliffs, whereas humid areas have rounded topography. Limestone, which is made of the soluble mineral calcite, is a resistant rock and forms bold outcrops in dry areas, but it is rapidly weathered and tends to form valleys in humid regions.

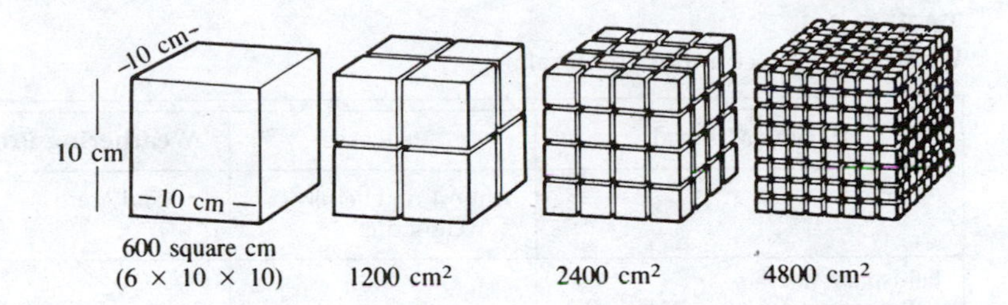

FIGURE 1.22
Surface area vs. particle size. Shows how mechanical weathering increases the surface area and thus helps to increase chemical weathering. A cube 10 cm on a side is shown in each case.

Chemical weathering can be defined as changes taking place near the surface that tend to restore the minerals to equilibrium with their surroundings. The primary minerals form, in most cases, at elevated temperature and pressure, and are in equilibrium under those conditions. When these minerals are exposed at the surface, they are in conditions far from those under which they formed, so they react to form new minerals that will be in equilibrium. One can then predict that the minerals formed at the highest temperatures will probably be affected first in chemical weathering. Thus it turns out that Bowen's reaction series (Fig. 1.10), which shows the sequence of mineral formation with falling temperature, also shows the relative stability of minerals to weathering, with olivine the most affected and quartz the least affected.

The products of chemical weathering of the common minerals are summarized in Table 1.5 (see also Fig. 1.23).

From this brief outline of chemical weathering, it is possible to make some interpretation of sedimentary rocks. As might be expected, most sandstones are composed largely of quartz because it is the mineral most resistant to weathering. Some sandstones contain much feldspar; such rocks formed under conditions that did not permit chemical weathering to complete its job. This situation could have developed several ways. Accumulation might have been too rapid to permit complete weathering, or the climate may have been too dry or too cold. Certainly there are other possibilities, but this serves as an example of how sedimentary rocks can be interpreted.

Soils Soils are of great economic importance. They can be said to be the bridge between terrestrial life and the inanimate world. **Soil,** as the term is used here, *is the material that supports plant life,* and so supports all terrestrial life. Only lichen and primitive plants, such as silver sword in Hawaii, can get nutrition directly from rock without soil. The development of soil is a complex interplay of weathering and biologic processes.

The importance of soil to life is shown by the composition of some foods. Steak contains about 12.5 percent phosphorus, iron, and calcium by weight. Thus, each pound of steak contains about two ounces of rock material obtained from soil by plants. A much larger amount of rock had to be weathered to release these two ounces. Plants use much weathered rock material. For example, an acre of alfalfa yields about four tons of alfalfa per year. To produce this alfalfa, more than two tons

TABLE 1.5
Weathering of common minerals.

Parent Mineral	Process	Weathering Product	Disposition
Quartz	Almost no chemical weathering	Sand grains	Sandstone
Feldspars Potassium feldspar ($KAlSi_3O_8$) K (potassium)	+ CO_2(carbonation)	→ K_2CO_3	Some is transported to ocean, some is used by plant life, and some is adsorbed on or taken into certain clays
Al, Si, O	+ H_2O (hydration)	→ Clay ($(Al_2Si_2O_5(OH)_4)$)	Shale
SiO_2		→ Soluble and colloidal silica	Chert (chalcedony)
Plagioclase feldspars Ca	+ CO_2	→ Calcite ($CaCO_3$)	Limestone
Na	+ CO_2	→ Na_2CO_3	Dissolves in ocean
Al, Si, O, and SiO_2	as above		
Muscovite—produces the same products as potassium feldspar			
Ferromagnesian minerals Depending on composition, form the same weathering products as the feldspars and also, Fe	+ O (oxidation)	→ Hematite (Fe_2O_3)	Sedimentary deposits of iron ore
	+ O + H_2O (oxidation and hydration)	→ Limonite ($FeO(OH)$)	Sedimentary deposits of iron ore
Mg	+ CO_2 (not clearly known)	→ $MgCO_3$	Some replaces calcium in limestone to form dolomite and some goes into certain clay minerals.

of rock must be weathered to obtain the phosphorus, calcium, magnesium, potassium, and other elements in the alfalfa (Keller, 1966). This corresponds to about 0.178 mm (0.007 inch) of rock weathered. However, the average rate of erosion is estimated to be about 2.5 cm (1 inch) per 9,000 years, so the alfalfa needs about 5.5 times the annual erosion rate. This "uses" soil at a very high rate, so fertilizers must be added. Thus soils can be used faster than they form, a fact not always taken into account in agricultural economics.

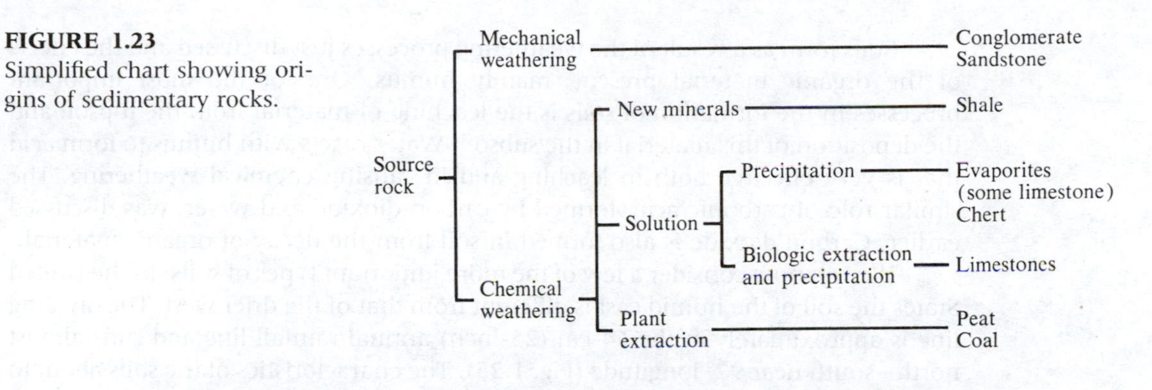

FIGURE 1.23
Simplified chart showing origins of sedimentary rocks.

There are many different types of soils. At first it was thought that the parent rock determined the soil type. We now know that climate is more important than parent rock; but the situation is complex, and other factors are involved, especially the type of vegetation. Soils can be described as mature and immature. On hillslopes, erosion may prevent soils from reaching maturity and the transported soil may accumulate in valleys (Fig. 1.24B). Many soils form on surface materials that have been transported. It apparently takes at least a few hundred years for a soil to develop. For this reason soil conservation is very important. The range in rate of soil formation is great. In 45 years, 35.6 cm (14 inches) of soil formed on pumice from the eruption of the volcano Krakatoa. At the other extreme, in many places no soil was formed on glacially polished surfaces in about 10,000 years.

In most soils three layers, or **horizons,** can be recognized. They are *the topsoil, the subsoil, and the partly decomposed bedrock.* In addition, in some areas, especially forests, a layer of organic material may form at the surface (Fig. 1.24). The material that gives the upper parts of many soils their dark color is humus. **Humus** *is formed by the action of bacteria and molds on the plant material in the soil.* Chemical weathering releases some of the materials used by plants, and humus also provides food for plants.

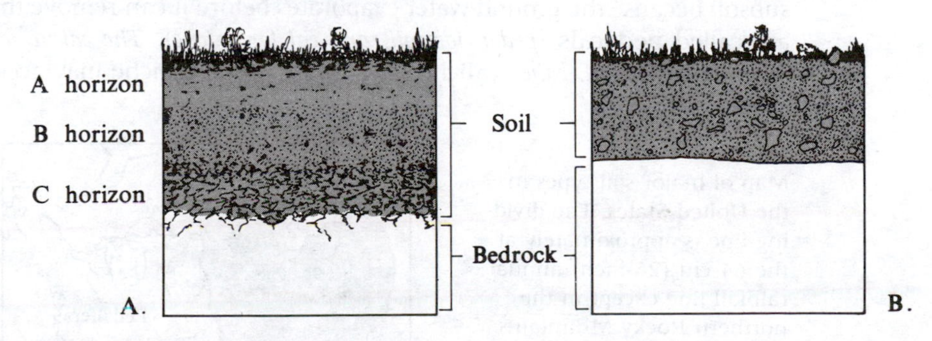

FIGURE 1.24
Soil profiles. A. Residual soil with well-developed horizons or layers. B. Tranported soil without layering.

Soils form as a result of the weathering processes just discussed and the effects of the organic material present, mainly humus. One of the most important processes in the formation of soils is the leaching of material from the topsoil and the deposition of this material in the subsoil. Water reacts with humus to form acid that is very effective both in leaching and in causing chemical weathering. The similar role of carbonic acid, formed by carbon dioxide and water, was discussed earlier. Carbon dioxide is also formed in soil from the decay of organic material.

We can now consider a few of the more important types of soils. In the United States the soil of the humid east is different from that of the drier west. The dividing line is approximately at the 64-cm (25-inch) annual rainfall line and runs almost north–south near 97° longitude (Fig. 1.25). The characteristics of the soils about to be discussed are summarized in Table 1.6.

The soil that predominates in the eastern United States is called **pedalfer** *(pedon, Greek for soil, plus Al, the symbol for aluminum, and fer for iron).* Humus is well developed in the temperate, moist climate. The abundant rain water becomes strongly acidic by reaction with the humus and leaches the topsoil. Aluminum and iron leached from the topsoil are deposited in the subsoil, giving it a brown (limonite) color. Soluble materials such as calcite or dolomite are leached from this type of soil and are generally removed by the ground water. These soils are quite acid, and clays are well developed. Ground water carries the clay to the subsoil so that topsoil may be somewhat sandy, due to resistant quartz fragments.

Pedalfers generally are best developed under forests. Conifer forests produce a thick layer of litter on the forest floor and so develop more humus. Therefore, the soil is very acidic and its development is accentuated; the resulting soil is light gray. It is in the pedalfers that calcite, generally called "agricultural lime" or "limestone," is spread on the soil to combat its acidity.

In the drier west, less humus develops, so the limited amounts of ground water are much less active chemically. Chemical weathering is also much slower in this climate, so there is less clay than in pedalfers. Under these conditions leaching is not complete, and the soluble materials, especially calcite, are deposited in the subsoil because the ground water evaporates before it can remove them. Such soils are called **pedocals** *(pedon for soil, and cal for calcite). The calcite forms as whitish material in the soil, called* **caliche** *(Spanish for lime).* Caliche may accumulate in the

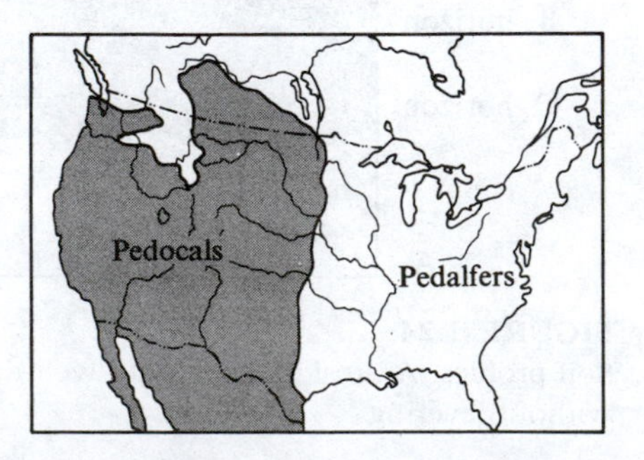

FIGURE 1.25
Map of major soil types in the United States. The dividing line is approximately at the 64-cm (25-inch) annual rainfall line except in the northern Rocky Mountains where low temperature affects the soil type.

TABLE 1.6
Generalized summary of soil types. See text for details.

Climate	Temperate humid [>64-cm (25-in.) rainfall]	Temperate dry [<64-cm (25-in.) rainfall]	Tropical savanna (heavy seasonal rainfall)	Extreme arctic, desert (cold or hot)
Vegetation	Forest	Grass-brush	Grass and tree	Almost none, so no humus develops.
Typical Area	Eastern U.S.	Western U.S.	—	—
Soil Type	Pedalfer	Pedocal	Laterite	
Topsoil	Sandy; light colored; acid.	Commonly enriched in calcite; whitish color.	Zones not developed. Enriched in iron (and aluminum); brick red color. All other elements removed by leaching.	No real soil forms because there is no organic material. Chemical weathering is very slow.
Subsoil	Enriched in aluminum, iron, clay; brown color (limonite).	Enriched in calcite; whitish color.		
Remarks	Extreme development in conifer forests because abundant humus makes ground water very acid. Produces a light gray soil because of removal of iron.	*Caliche* is the name applied to the calcite-rich soils.	Apparently bacteria destroy the humus so no acid is available to remove iron. Tropical rain forests develop soils similar to pedalfers.	—

topsoil as well as in the subsoil. This can occur when the ground water moves upward through capillary openings because of the evaporation at the surface. Pedocals tend to develop under grass- or brush-covered areas.

Soil formation in tropical climates is not fully understood, in part because few areas have been well studied. The soil in rain forests is somewhat similar to the pedalfers. The soil of the grass- and tree-covered savannas, however, is much different. Presumably because of the seasonal heavy rainfall, leaching is the dominant process. Almost everything, including silica and clays, is leached from the soil and carried away by ground water except iron and aluminum, which form hydrated oxides. Apparently, at the high temperatures, bacteria destroy the humus so that the ground water is not acidic and cannot remove iron and aluminum. The soils produced in this way are called laterites (Fig. 1.26).

Laterite is from the Latin word for brick, and its reddish-brown color is similar to that of a brick. Laterites do not form distinct layers as other soils do. As might be expected, they are poor soils on which to grow crops, as shown by the many unsuccessful attempts to develop new farmlands in emerging tropical

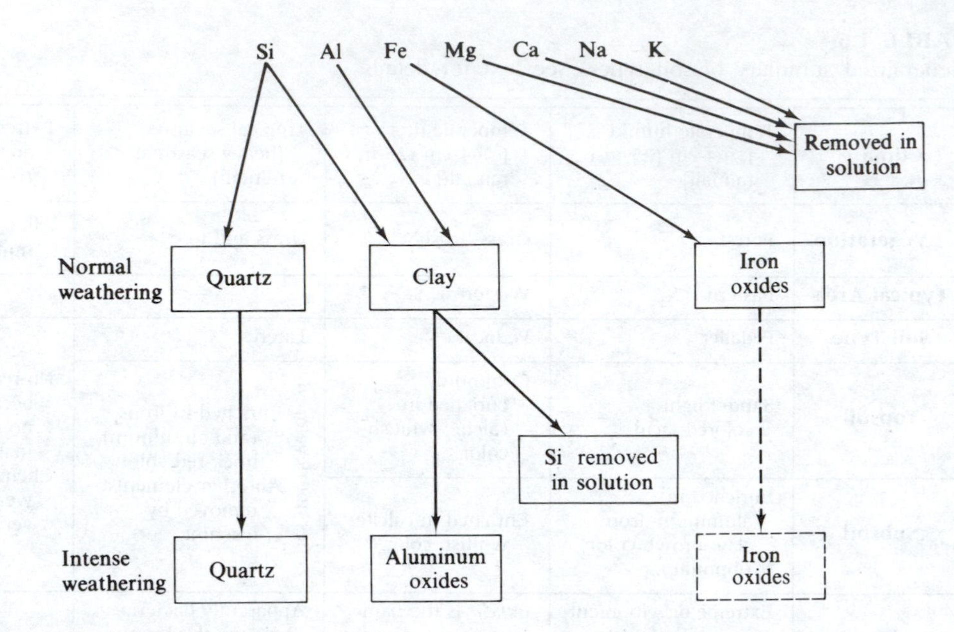

FIGURE 1.26
Simplified summary of weathering, showing how laterites form.

countries. Most laterites are iron-rich; however, in some areas they are mainly aluminum. Such *a mixture of hydrated aluminum oxides is called* **bauxite** and is a valuable ore. Bauxite probably forms only from the weathering of aluminum-rich rocks that have little iron, such as some feldspar-rich igneous rocks. At some places laterites are mined for iron ore and, at a few places, for their nickel or manganese; here the parent rock probably was rich in these elements.

In the extreme climates, such as arctic and desert, very little chemical weathering occurs and there is almost no organic material; therefore, no real soil can develop.

Lithification *The transformation of a sediment into a rock is called* **lithification.** Several processes are involved, including cementation, compaction, and recrystallization.

Cementation *is the deposition by ground water of soluble material between the sediment grains* and, as might be expected, is most effective in coarse-grained, well sorted, permeable rocks. The main cementing agents are:

Calcite—recognized by acid test (do not mistake this effervescence of the cement for the overall effervescence of a calcite-bearing rock.)
Silica—generally produces the toughest rocks
Iron oxides—color the rock red or yellow

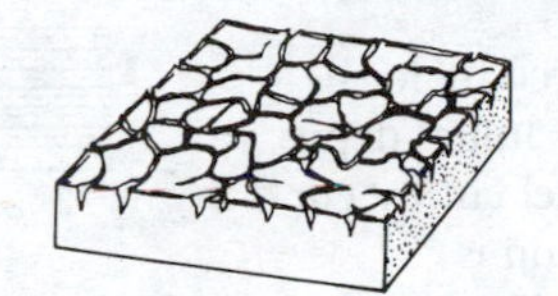

FIGURE 1.27
Mud cracks, which form as a result of drying mud flats, can be used to determine top and bottom.

Other cements such as dolomite are possible. In certain poorly sorted rocks, clay, which generally colors the rock gray or green-gray, may be thought of as forming the cement.

Compaction is an effective lithifier of fine-grained rocks such as shale and siltstone. *Compaction generally comes about from weight of overburden* during burial, and the reduction in volume due to compaction and squeezing out of water may amount to more than 50 percent in some shales.

Recrystallization probably is important in producing chert and in transforming calcite-bearing mud into limestone. Recrystallization produces an interlocking texture.

Features of Sedimentary Rocks The most noticeable feature of an outcrop of sedimentary rocks is the **bedding,** which records the layers in the order of deposition with the oldest at the bottom. In some instances the beds are too thick to show in a small outcrop, and, for the same reason, many hand specimens do not show obvious bedding.

Sedimentary rocks may contain fossils and sedimentary features which can tell much about the environment of deposition. **Mud cracks** on bedding planes record periodic drying; they suggest shallow water and perhaps seasonal drying (Fig. 1.27). **Ripple marks** also suggest shallow water with some current action, but are also known to form in deep water. Detailed study of ripple marks can show type and direction of current (Fig. 1.28). Current action can also cause localized scouring or cutting to produce **cut-and-fill** features (Fig. 1.29). Another type of current feature is **cross-bedding,** in which a depression is filled by slanted beds (Fig. 1.30). These features can also be used to recognize beds that have been overturned by folding or faulting (Fig. 1.31).

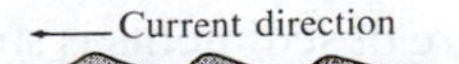

FIGURE 1.28
Ripple marks form where currents can act on sediments. To see why only wave ripple marks can be used to determine top and bottom of beds, turn the page upside-down.

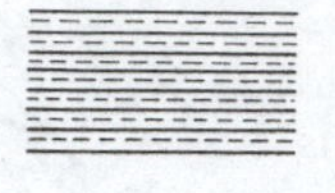

FIGURE 1.29
Scouring produced by localized currents. A. Initial deposition. B. Channel cut by current. C. Deposition is resumed, filling the channel. Such a feature records an interruption in deposition and can be used to distinguish top and bottom.

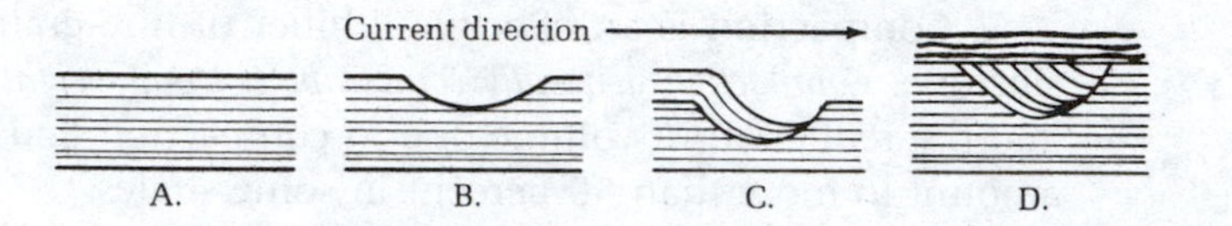

FIGURE 1.30
Development of cross-beds. A. Initial deposition. B. Channel cut by a current. C. Channel filled by deposition from one side. D. Normal deposition after channel is filled. Cross-beds can also be used to distinguish top and bottom.

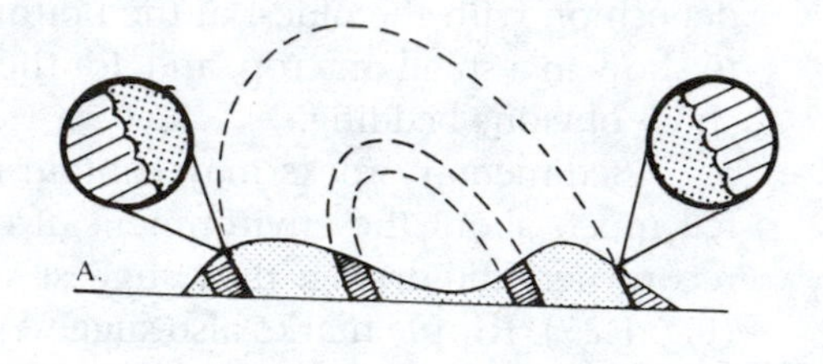

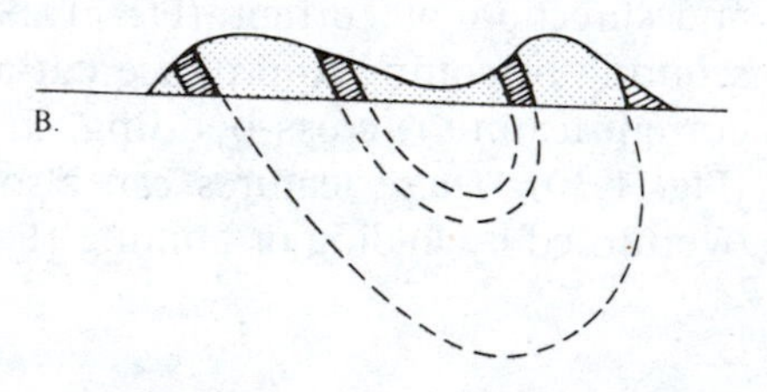

FIGURE 1.31
Alternate interpretations of structure from outcrops of folded sedimentary rocks. The ripple marks in A show that it is the correct interpretation. In the absence of ripple marks, or other sedimentary features that indicate top and bottom, it would not be possible to choose between the two interpretations.

If a mixture of sediments is suddenly deposited into a sedimentary basin, then the large fragments sink faster than the small ones. This does not violate Galileo's famous gravity experiment, because here we are concerned with falling bodies in a viscous medium in which velocity of descent is controlled mainly by the size of the body. The type of bedding produced by this kind of sedimentation is called **graded bedding** (Fig. 1.32). We can demonstrate this experimentally by putting sediments of various sizes in a jar of water, shaking, and then allowing the contents to settle.

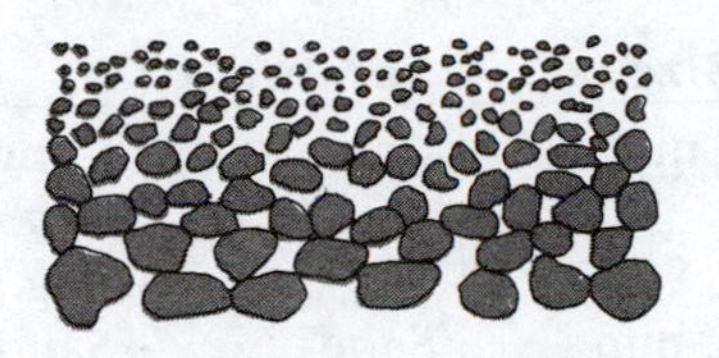

FIGURE 1.32
Graded bedding. Produced by the more rapid settling of coarse material than fine material.

The **sorting** *of a sediment is a measure of the range in size of its fragments.* If the fragments that compose a sediment are all similar in size, the sediment is well sorted. The sorting of a sedimentary rock determines other textural features. The **porosity** *of a sedimentary rock is a measure of its empty or void space,* and its **permeability** *is a measure of the interconnections of the pore spaces.* These properties become quite important when one is concerned with recovering oil, natural gas, or ground water from an underground reservoir.

The amount of pore space in sedimentary rocks is quite varied but surprisingly large. If we consider packing of uniform spheres (most sand grains are subspherical), we find that the closest possible packing contains 26 percent void space, and the loosest, 48 percent (Fig. 1.33). The permeability varies depending on the amount of pore space, the sorting, the cementation, and the size of the particles. A shale may have over 30 percent porosity, but the surface tension of water will prevent flowage through the tiny openings.

Classification As with the igneous rocks, it is much more important to interpret the origin of sedimentary rocks than merely to name them. Knowing the source of the materials that compose a sedimentary rock and understanding the cause of its sedimentary features will enable such interpretation.

According to the way they were formed, sedimentary rocks can be classified as marine, lacustrine (lake deposited), glacial, eolian (wind deposited), or fluvial (river deposited). They can be classified also by composition as limestone, chert, quartzose sandstone, etc., or by their mode of origin as **clastic,** chemical precipitate, or organic. In practice, these are blended to produce a practical classification as follows.

Clastic rocks **Clastic rocks** *are composed of rock fragments or mineral grains broken from any type of preexisting rock. They are subdivided according to fragment size.* Commonly, sizes are mixed, requiring intermediate names such as sandy siltstone. They are recognized by their clastic texture (Fig. 1.34).

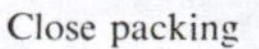

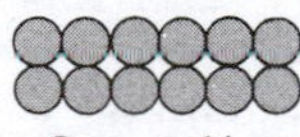

FIGURE 1.33
The packing controls the porosity, as illustrated here in the idealized case of spheres. Compare with Figure 2.44.

Size	Sediment	Clastic Rock
Over 2 mm	Gravel	**Conglomerate**—generally has a sandy matrix (Fig. 1.13G). **Breccia** has angular fragments (Fig. 1.13H).
1/16–2 mm	Sand	**Sandstone**—recognized by gritty feel (Fig. 1.13F). Generally designated as coarse, medium, or fine if well sorted. Sandstones have also been subdivided on the basis of composition. The most important types are: *Quartzose sandstone* (mainly quartz) *Feldspathic sandstone* (arkose)—over 20% feldspar *Graywacke*—poorly sorted, with clay or chloritic matrix.
1/256–1/16 mm	Silt	**Siltstone (mudstone)**—may be necessary to rub on teeth to detect grittiness, thus distinguishing from shale.
Less than 1/256 mm	Clay	**Shale**—distinguished from siltstone by its lack of grittiness and its *fissility* (ability to split very easily on bedding planes).

Nonclastic Rocks **Nonclastic rocks** *are formed by chemical precipitation, by biologic precipitation, and by accumulation of organic material. They are classified by composition.* As with the clastic rocks, these are commonly mixed both among themselves and with the clastic rocks.

Limestone is composed of calcite. Recognized by effervescence with dilute hydrochloric acid. Generally of biologic origin and may contain fossils (Fig. 1.13I). A rock composed mainly of fossils or fossil fragments is called *coquina.*

Dolomite is composed of the mineral dolomite. After scratching (to produce powder), recognized by effervescence with cold dilute hydrochloric acid; will also react (without scratching) with concentrated or with warm dilute hydrochloric acid. Generally formed by replacement of calcite, presumably very soon after burial. The reduction in volume in this replacement may produce irregular voids and generally obliterates fossils.

Chert is composed of chalcedony and is identified as discussed under mineralogy. The names chert and chalcedony were applied prior to the advent of

FIGURE 1.34
Recognition of textures is very important in the identification and interpretation of rocks.

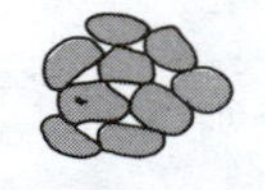
Clastic texture

Interlocking texture. Developed in chemical precipitates, such as some limestones, and in many igneous and metamorphic rocks.

the microscope and the X-ray, and were based on minor color and luster differences that cannot be substantiated by modern methods. Dozens of names have been applied to chalcedonic silica; of these, a few may be useful, e.g., *agate* for banded types, and *flint* for dark gray or black chert. Chert may form either beds or nodules.

Rock salt is composed of halite and may be recognized by its cubic cleavage and its taste (see mineral tables). It is generally deposited with other salts termed **evaporites** because they form when restricted parts of the sea are evaporated.

Chalk is soft, white limestone formed by the accumulation of the shells of microscopic animals. Recognized by effervescence with acid.

Diatomite is a soft, white rock composed of the remains of microscopic organisms (diatoms). Because the remains are composed of silica, it is distinguished from chalk by lack of effervescence.

The rocks described above are the most important, but by no means the only, sedimentary rocks. Any mixture of types is possible and, because any weathering product may form a sedimentary rock, endless variety is possible. Some less abundant types include economically important deposits such as coal, iron oxides, phosphorous rocks, bauxite (aluminum ore), and potash.

Shale, sandstone, and limestone make up about 99 percent of all sedimentary rocks. Because feldspars are the most abundant minerals, their main weathering product, shale, is the most abundant sedimentary rock. Sandstone is next in abundance, and then limestone.

Sedimentary Rocks and Plate Tectonics It was noted long ago by geologists that many mountain ranges are formed of bent and broken layers of sedimentary rocks. These sedimentary rocks had been deposited as flat layers in shallow seas. It is known that the seas were shallow because the rocks have features that commonly form in shallow water, such as mud cracks, and the fossils they contain are organisms that live in shallow waters. The mountain ranges, then, began as elongated basins of deposition, and these areas of sedimentation were named **geosynclines.** The sedimentary rocks in geosynclines are much thicker than nearby rocks of similar age. Geosynclinal mountain ranges are commonly near the margins of continents, and so some of them may have formed in the continental shelf–continental slope area.

The sediments of **continental shelves** tend to be sandstones near the coast, with shale or limestone accumulating farther from shore, although many exceptions are found (Fig. 1.35). The rocks that underlie the continental slope are generally clastics, shale and sandstone, commonly intermixed. Few or no volcanic rocks are found in the continental shelf–continental slope area.

In most geosynclinal mountain ranges, volcanic rocks are found interbedded with sedimentary rocks in some areas. This suggests that in these cases, convergent plate boundaries may be involved. Volcanic islands form where oceanic plates collide with continental plates and so are also at the margins of continents (Fig. 1.36). The volcanic borderlands can be the source of both volcanic rocks and clastic sediments. Geosynclinal clastic sediments containing volcanic rocks generally are

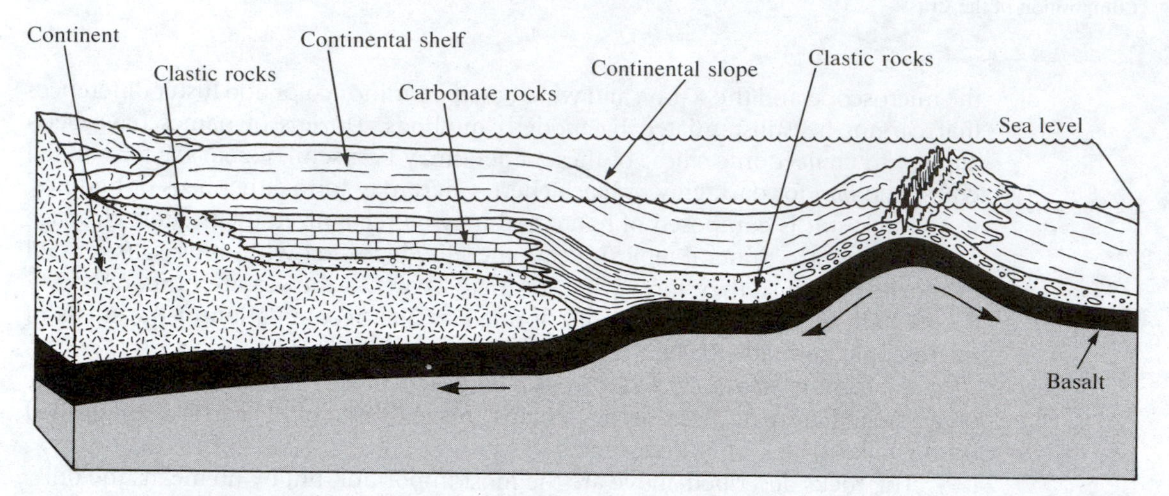

FIGURE 1.35
Development of thick sections of sedimentary rocks at a continental shelf and slope, and much thinner layers of sediments and volcanic rocks on the ocean floors and at mid-ocean ridges.

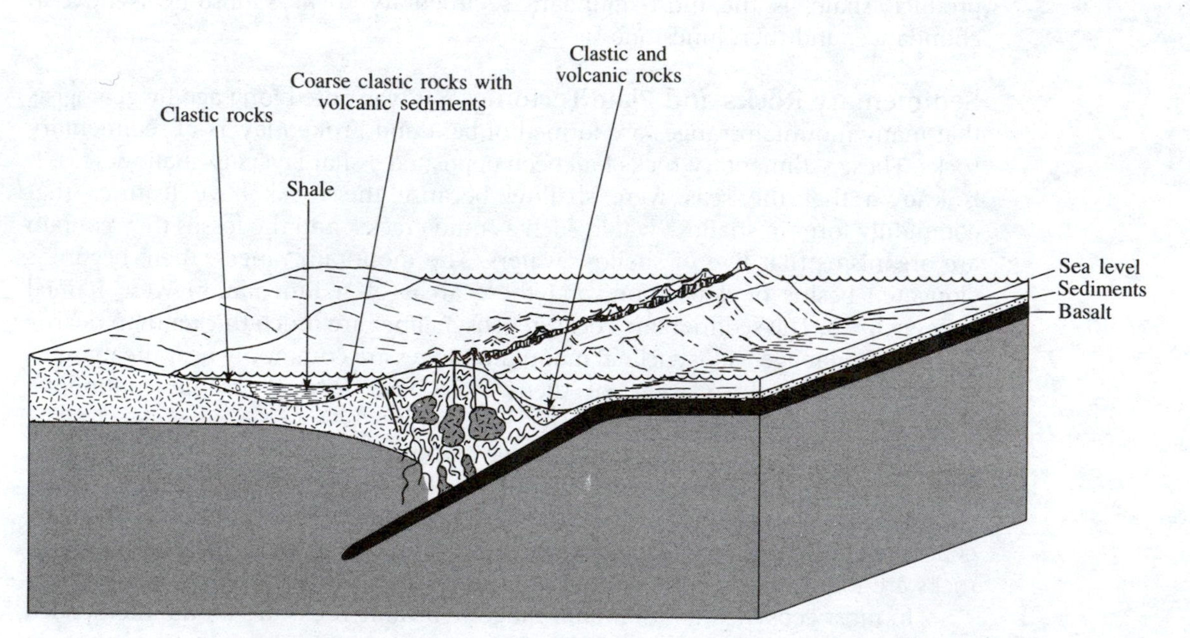

FIGURE 1.36
Development of thick sections of sedimentary and volcanic sedimentary rocks at a convergent plate boundary.

very poorly sorted and therefore have a wide variety of sizes and types of fragments. Such rocks contrast with the sediments on continental shelves, which tend to be well sorted. Later plate movements are believed to crumple the geosynclinal sediments.

Metamorphic Rocks

Metamorphic rocks *are rocks that have been changed either in texture or in mineral composition by any of the following—heat, pressure, stress (directed pressure), shear, or chemically active solutions.* Figure 1.37 illustrates some of these. Note that the changes that take place in weathering and in the lithification of a sedimentary rock are metamorphic, according to this definition. Such changes, however, are not considered metamorphic, but are arbitrarily excluded.

Types of Metamorphism Metamorphic rocks can be subdivided into three genetic groups, although there is gradation among the groups. They are:

1. **Thermal or contact metamorphic rocks,** *generally found at the margins of intrusive igneous bodies such as batholiths.* Such rock is termed **hornfels.**
2. **Regional metamorphic rocks,** *so-called because they generally occupy large areas.* These rocks form deep in the crust, and their presence at the surface reveals much uplift and erosion. *The rocks generally have recrystallized under stress so that the new minerals grow in preferred orientation.* These rocks are said to be **foliated,** and most are **gneiss** (pronounced "nice") and **schist.**

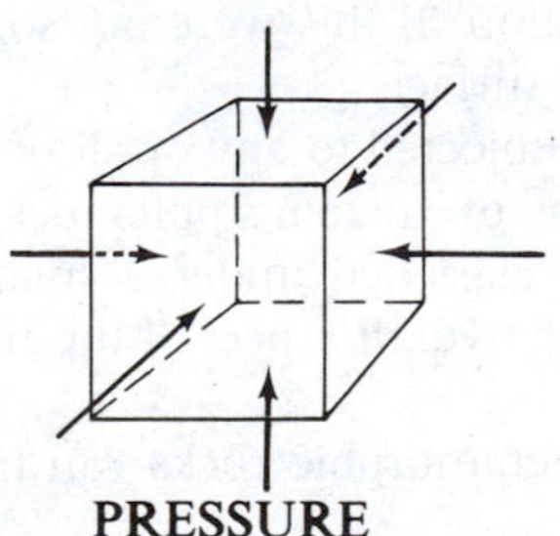

PRESSURE

Force uniform on all sides

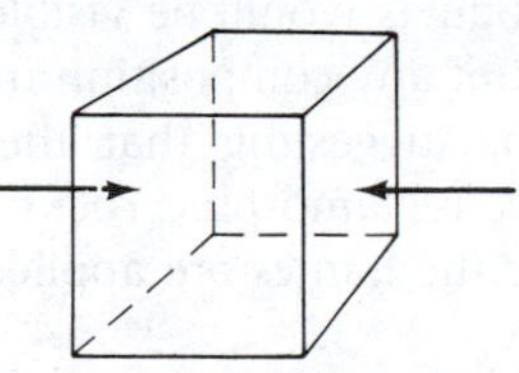

STRESS (directed pressure)

Force uniform on two opposite sides

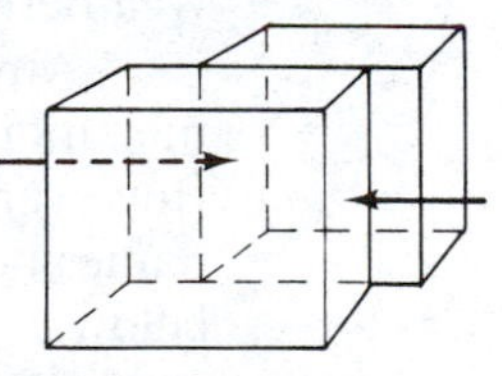

SHEAR (breakage)

Force at different places on opposite sides can cause deformation or rupture

FIGURE 1.37
Pressure, stress, and shear.

3. **Dynamic metamorphic rocks,** *resulting from breaking and grinding without much recrystallization.* **Mylonite** is a rock formed in this way.

Study of metamorphic rocks, coupled with laboratory experiments, yields much information about conditions and processes deep in the crust. Thus, if we know the conditions of formation of a mineral (or better, an assemblage of minerals), then when we find such minerals in rocks, we know the conditions under which the rocks formed. However, the problem is complicated by several factors: the diversity of natural rocks, which contain many more elements than can be easily studied in experiments; the impossibility of reproducing the effects of geologic time in the laboratory; and the difficulty of detecting the effects of solutions or volatiles that may have catalyzed the reactions in natural rocks. Thus, although we are very rapidly gaining an understanding of metamorphic processes, many problems remain.

Types of Metamorphic Rocks The purpose of this section is to give the student some ability in identifying metamorphic rocks. One will not, except in a very general way, be able to interpret the detailed geologic history of these rocks without knowledge of chemistry, mineralogy, and field and microscope study. To interpret metamorphic rocks fully, one must know whether they contain equilibrium mineral assemblages or whether conditions changed before the metamorphic reactions were complete; whether material was added or subtracted during metamorphism; and whether the rocks were subjected to one or more periods of metamorphism. One can think of many more possibilities that must be considered; for example, if the reactions take place during deep burial, will the rocks revert to something like their original state during slow uplift? No doubt some such effect may operate in some instances, but most of the chemical reactions involved go very slowly in the reverse direction. If this were not so, nothing but weathering products would be visible at the surface.

Any rock of any composition may be subjected to any or all of the agents of metamorphism, suggesting that the variety of metamorphic rocks is endless. However, most metamorphic rocks can be classified under a small number of names because the names are applied for the overall aspect of the rocks as noted below.

For classification and identification, metamorphic rocks can be subdivided into two textural groups:

1. **Foliated**—*having a directional or layered aspect.*
2. **Nonfoliated**—*homogeneous or massive rocks.*

Nonfoliated Rocks Because the nonfoliated rocks are the simpler of the two types, we will consider them first. There are two types of nonfoliated metamorphic rocks. The first consists of *thermal or contact metamorphic rocks called* **hornfels** (singular). They generally occur as narrow belts around intrusive bodies and may originate from any type of parent rock. Sometimes these are called *baked rocks;* the name is appropriate because their formation is quite similar to the baking of clay pottery in a kiln. They are generally fine-grained, tough rocks that are difficult to identify without microscopic study unless the field relations are clear. They range from being completely recrystallized, with none of their original features preserved, to being slightly modified rocks, with most of the original features preserved. Calcite-bearing contact rocks are generally quite spectacular, containing large crystals of garnet and other minerals.

The second type of nonfoliated metamorphic rock develops if the newly formed metamorphic minerals are equidimensional and so do not grow in any preferred orientation. Examples are the rocks that result when monomineralic rocks such as limestone, quartz sandstone, and dunite are metamorphosed. In the case of limestone, which is composed of calcite, no new mineral can form; thus, the *calcite crystals, which are small in limestones, grow bigger, develop an interlocking texture, and become* **marble** (Fig. 1.13J). Marble can be distinguished from limestone only by its larger crystals and lack of fossils; being composed of calcite, both effervesce in acid. Marble may have any color; pure white calcite rocks are more apt to be marble than limestone. Impure limestones, as noted above, develop new minerals, particularly garnet. *Dolomite marble also occurs.*

Because quartzose sandstone similarly can form no new minerals, the quartz crystals enlarge and intergrow to form **quartzite.** Quartzite is distinguished from chalcedony, chert, and opal by its peculiar sugary luster. The term quartzite is best reserved for the metamorphic rock, although quartz sandstone is called quartzite by some, and the metamorphic rock is called metaquartzite.

Both marble and quartzite may form under stress, so the crystals may be aligned. However, because the crystals are equidimensional, this preferred orientation can be detected only by microscopic examination. Thus, although we call these rocks nonfoliated, some may have a hidden foliation.

So far we have been concerned with **progressive metamorphism,** *which is metamorphism caused by increased temperature, pressure, and other metamorphic agents.* The next example is one of **retrogressive metamorphism,** *which is metamorphism at a temperature lower than that of the rock's original formation.*

Dunite is composed mainly of olivine, which, in Bowen's series, is the mineral formed at the highest temperature. When it is subjected to the same metamorphic conditions that produce the rocks discussed earlier, it too is metamorphosed—but at a temperature lower than that of its formation. The olivine of dunite is changed to the mineral *serpentine,* and the metamorphic rock produced is called **serpentinite.** Its recognition is discussed in the mineral tables. In other instances, instead of serpentine, *talc* is the metamorphic mineral that forms. If nonfoliated, the talc rock is called **soapstone;** if foliated, it is called **talc schist.** Identification of talc is also discussed in the mineral tables. Slight differences in the composition of the parent rock or of the fluids that cause this type of metamorphism probably control which mineral is formed.

Foliated Rocks Regional metamorphism generally takes place in areas that have undergone severe deformation. These very active areas are generally the cores of mountain ranges. The same forces that fold and fault the shallow rocks exposed on the flanks of mountain ranges provide the stress fields under which the deep-seated metamorphic rocks exposed in the uplifted cores of mountain ranges were recrystallized. An example of the rocks produced by increasing metamorphic grade will show most of the foliated metamorphic rock types. These same rock types will form when many other parent rocks are metamorphosed; the differences in original chemical composition will change only somewhat the relative amounts of the metamorphic minerals.

Shale is composed mainly of clay. During lower-grade metamorphism, the clay is transformed to mica, and at higher grades of metamorphism, to feldspar. Which micas and which feldspars form depends on the bulk chemical composition. Compare this with chemical weathering and note that weathering is, in a sense, retrogressive metamorphism. Because mica is a flat, platy mineral, it grows with its leaves perpendicular to the maximum stress, forming a preferred orientation (Fig. 1.38).

The sequence produced in the metamorphism of a shale is as follows:

Sedimentary rock	*Low-grade metamorphism*	*Medium-grade metamorphism*	*High-grade metamorphism*
SHALE ⟶	SLATE ⟶	SCHIST ⟶	GNEISS
Clay	Clay begins to be transformed into mica. Mica crystals are too small to see, but impart a cleavage to the rock. May also form by mechanical rearrangement alone.	Mica grains are larger so that the rock has a conspicuous foliation.	Mica has transformed largely to feldspar, giving the rock a banded or layered aspect.

The rock names are applied for the overall *texture* of the rock and not strictly for the mineral transformations, which occur over a range. Most schists contain

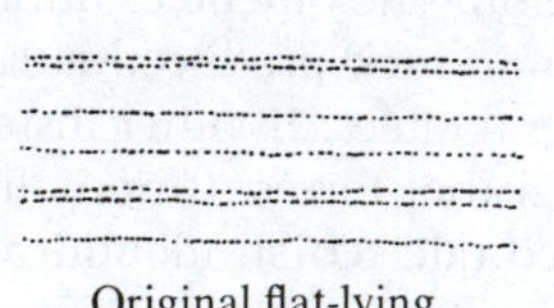

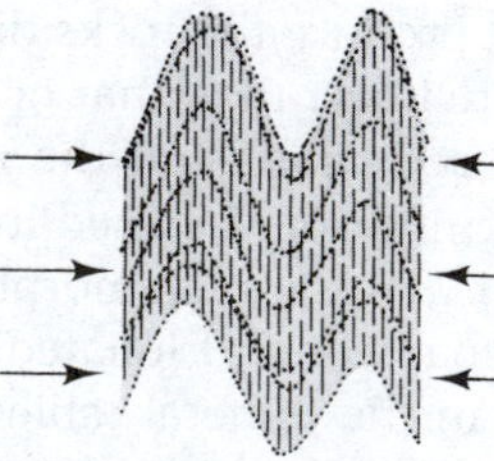

FIGURE 1.38
Development of foliation.

feldspars, but the name gneiss is reserved for rocks with much more feldspar than mica (Figs. 1.13K and L, 1.39, and 1.40). Other rocks that have the same general bulk chemical composition, thus producing the same metamorphic rocks, are certain pyroclastic and volcanic sedimentary rocks, many sandstones, arkose, granite, and rhyolite. The coarse-grained rocks in this list are rarely the parents of

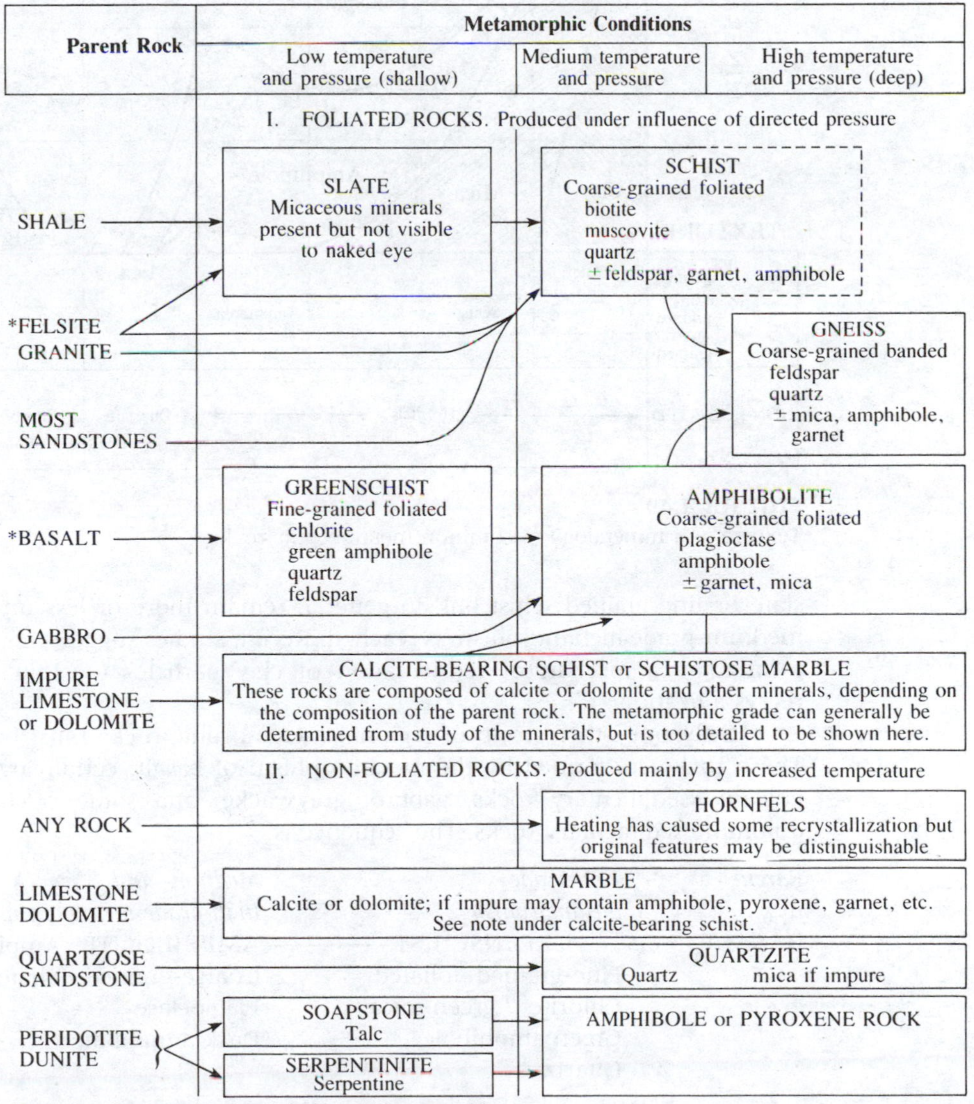

FIGURE 1.39
A generalized chart showing the origin of common metamorphic rocks.

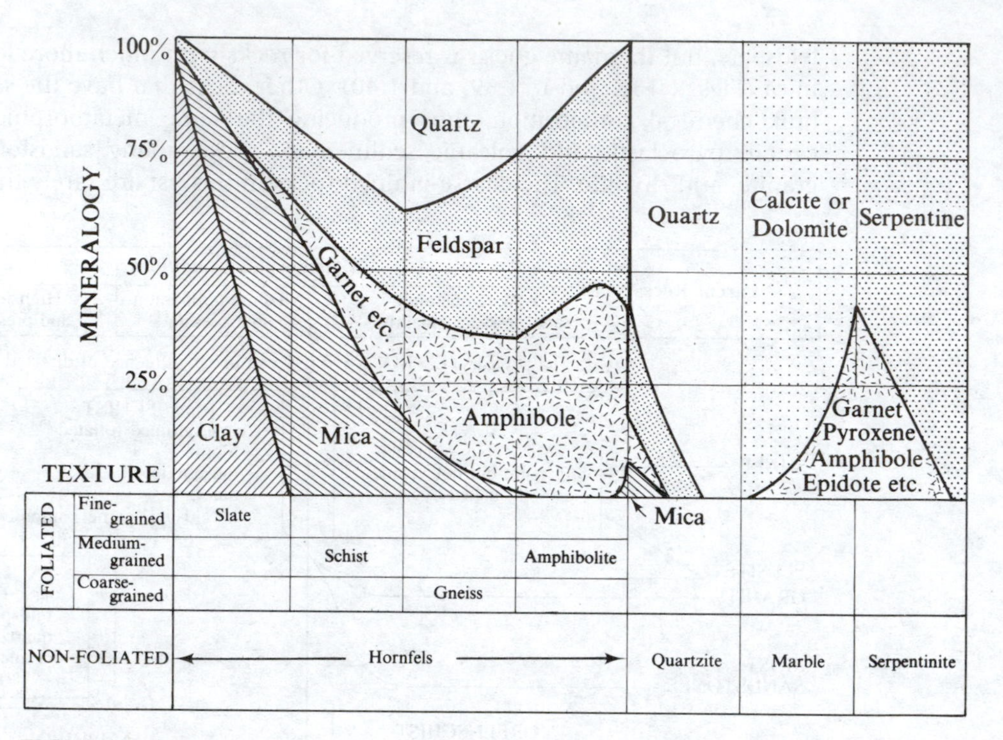

FIGURE 1.40
Texture and mineralogy of common metamorphic rocks.

slate or fine-grained schist but, in general, remain more or less unaffected until medium-grade metamorphism is reached. Recent studies suggest that slate may be formed by a mechanical reorientation of clay particles together with limited recrystallization.

Only one other group of regional metamorphic rocks must be considered. These rocks are formed by the metamorphism of basalt, certain pyroclastic and volcanic sedimentary rocks, gabbro, graywacke, and some calcite-bearing or dolomitic sedimentary rocks. The sequence is:

Parent rock	*Low-grade metamorphism*	*Medium- and high-grade metamorphism*
BASALT ⟶	GREENSCHIST ⟶	AMPHIBOLITE (Amphibole schist)
	Fine-grained, foliated	Coarse-grained, foliated
	Chlorite ("green mica")	Plagioclase
	Green amphibole	Dark amphibole
	Quartz	

Dynamic metamorphism, *the third type of metamorphism, results when rocks are sheared, broken, and ground near the surface when the temperature and pressure are too* *low to cause any significant recrystallization.* Thus, these rocks are commonly associated with fault zones, but there are all gradations between these rocks and ordinary schist, especially lower-grade schist formed from coarse-grained rocks. The fine-grained, ground rock in a fault zone, sometimes called *gouge,* is *mylonite;* and the rocks gradational with schist are called *semischist* or *cataclastic schist* or *gneiss.* The origin of these rocks is clear when they are encountered in the field, but they are very difficult to identify in hand specimen.

Metamorphic Rocks and Plate Tectonics The active areas of plate tectonics—mid-ocean ridges and convergent plate boundaries—are also places where metamorphism occurs. At mid-ocean ridges, the rocks near the crest are metamorphosed, as shown in Figure 1.41. The crest is an area of active basalt volcanism, so it is not surprising that the nearby rocks are subjected to metamorphic conditions. Volatiles, especially water, from the volcanic activity and the heat that generates the magma are probably important causes of this metamorphism.

Where an oceanic plate collides and moves downward under a continental plate, the resulting igneous and metamorphic rocks are easier to study because they are above sea level. The most common metamorphic rocks are the regional metamorphic rocks, and they are formed at such places. The heat and volatiles that cause the metamorphism are apparently associated with the rising magma that forms the volcanoes and the batholiths, as shown in Figure 1.42.

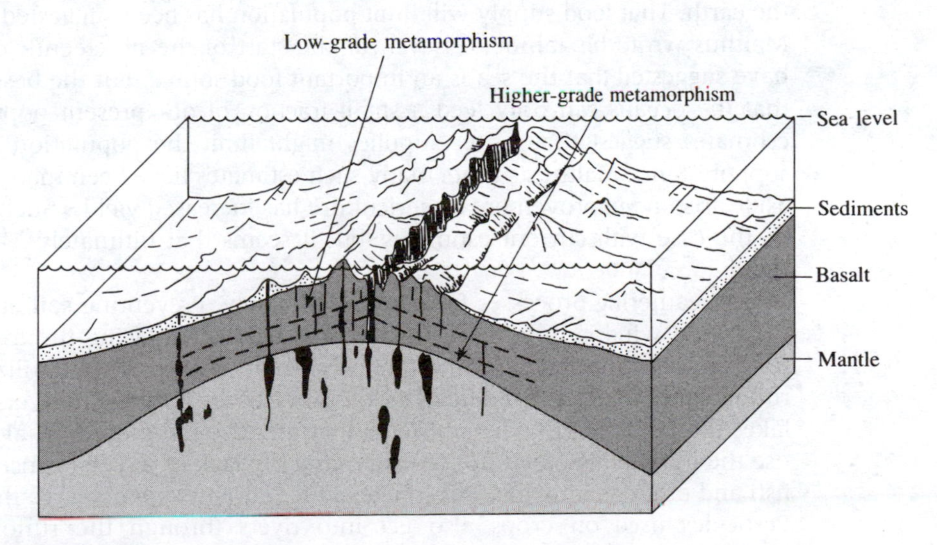

FIGURE 1.41
Metamorphism at a mid-ocean ridge.

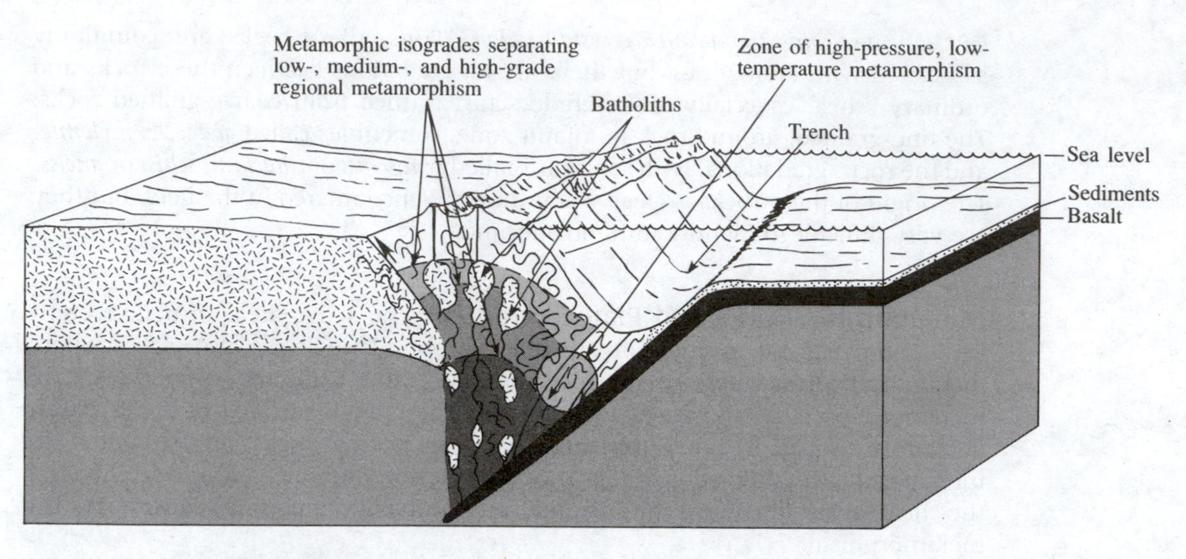

FIGURE 1.42
Metamorphism at a convergent plate boundary.

Resources

Water, Population, and Food Water and soil are the resources that ultimately support all life. Their availability will probably control the ultimate population of the earth. That food supply will limit population has been suggested since Thomas Malthus wrote his famous essay near the start of the nineteenth century. Some have suggested that the sea is an important food source, but the best estimates are that the oceans can only feed a small fraction of our present population. Some estimates suggest that water supplies might limit the population the earth can support. Since Malthus's time, many such estimates have been made and then cast aside as some improvement in agriculture has increased yields. Such will probably be the case with current estimates, but it seems that ultimately Malthus will be right.

Weathering processes form soil very slowly. Preventing soil erosion by both water and wind is important to food production. Replacing the materials in soil used by crops is also very important. A current problem with fertilizers is that the runoff water from fertilized fields carries some of the fertilizer to rivers. In rivers and lakes the fertilizer provides nutrients that increase the growth of algae. The algae use the oxygen dissolved in the water, and the lack of oxygen causes the death of fish and other aquatic life. Phosphates in laundry detergents have the same effect. Pesticides used on crops also get into rivers through the runoff. Long-lived insecticides such as DDT got into most types of life in this way, and via animals eating poisoned insects. Thus, keeping soil productive without causing harm to the environment is a problem that requires much study.

Mineral Resources The mineral resources of the earth are finite, and, for example, once mined, no more new ones can be had. This is one aspect of the spaceship earth analogy. The day when any one mineral runs out can be put off by improving recovery technology so that poorer and poorer grades of material can be mined, by recycling old metal, and by developing substitutes. Some metals, such as aluminum, will probably never run out (see Table 1.1), but some others, such as silver and mercury, may very well be exhausted.

No country is self-sufficient in all mineral resources, and the industrial countries are using mineral resources in much greater quantities than are the underdeveloped countries. The underdeveloped or nonindustrial countries are striving to become industrial and raise their standards of living. Thus the rate of use of minerals is bound to increase, and the increase will be rapid because most of the earth's population lives in the underdeveloped countries.

Some examples of metal usage will show the magnitude of the problem. In the United States, the per capita usage of iron is about six times that of the world average, or said another way, iron production would have to be increased six times if the whole earth were to have the same standard of living as the United States. It is estimated that the population of the earth will double by the year 2000. To keep the same world per capita use of iron, production will have to be doubled; or if everyone is to have the United States's standard of living, production must be increased about 12 times. Of course, not only iron but all other mineral resources will be needed in similar ratios to achieve these goals. These projections may be too high for a number of reasons, such as export of manufactured goods by the United States. On the other hand, per capita consumption in the United States and elsewhere has been rapidly accelerating, suggesting that the projections may be too low. It seems clear that the potential demand for metals in the year 2000 will be at least several times that of today. The question is, can this potential demand be met?

The differences in mineral usage reveal a number of differences among countries. The United States is a highly industrialized country. With only 6 percent of the world's population and 6 percent of the earth's land surface, it uses about 16 percent of the earth's total production of coal, 25 percent of the petroleum, 26 percent of the steel, 35 percent of the copper, and 53 percent of the aluminum. It must be remembered, too, that these amounts add to the already large quantities of these metals that are now in use in buildings, machinery, and the like. The amounts of mineral resources used are increasing in all countries; but although the amounts used per capita are increasing in the industrial countries, they are decreasing per capita in many underdeveloped countries. The rapid growth of population in the underdeveloped countries is the cause. The industrial countries are increasing their standards of living, while in the underdeveloped countries, the people are becoming poorer, at least in mineral resources.

The United States, like most industrial countries, is not self-sufficient in many mineral resources. It must import large quantities. A question of great importance is, what is a fair price to pay an underdeveloped country for its unreplenishable mineral resources? All people look forward to the day when they will have a standard of living like ours. This will be difficult to achieve if their resources are gone. On the other hand, they cannot develop or even use their resources without capital, and their only ready way to get capital is to sell their resources. It takes many millions of dollars to find a resource, and many more to develop it, along with much technical know-how.

The historical background to the present world situation shows how underdeveloped countries have been exploited in the past. The colonial system flourished with the industrial revolution. The mother country was generally an industrial country, and the colonies supplied the mother country with raw materials. The political breakdown of this system began with the American Revolution and was nearly completed after World War II with the formation of many new independent countries in Africa and Asia. Most of the colonialism of today is economic colonialism and revolves around the question of the price paid for mineral resources. Prices may have to increase, and this may change the standards of living of the whole world. The constant upward revision of the royalties paid for oil in the Near East is an example of the kinds of changes that seem likely to occur. Anyone who does not believe the importance of mineral resources in world politics needs only to study the history of Europe for the past 200 years or so. The territorial settlements after each war involved mainly the important mineral resource areas, such as the iron of Alsace-Lorraine (France), the coal of Saarland (West Germany), and the lead-zinc of Poland.

Energy Resources Modern civilization depends on energy. At first our only use of energy was in the form of food. Later we used wood fires, and the industrial revolution required the use of coal. The fossil fuels, coal and petroleum, are the main energy sources at present, with nuclear power developing very rapidly.

The energy sources of the earth are:

- Solar radiation
  - Food
  - Fossil fuels
  - Water power
  - Wind power
  - Direct solar energy
- Tidal energy
- Geothermal energy
- Nuclear energy

The amount of energy that the earth receives from the sun is many times more than the amount we utilize. If a practical way to use it could be devised, no other source would be needed. Remember that the earth reradiates the energy received from the sun. If it did not, the temperature of the earth would increase, and geologic history shows that the average temperature has remained about the same for at least a half billion (500 million) years. Solar energy is our most important energy source. It is used by plants, providing the food that supports all life on earth. In the past, plants in the form of firewood provided an important source of energy. A very small amount of solar energy is preserved in the form of fossil fuels such as coal and petroleum.

Although the fossil fuels are our present main source of energy, their future is not bright. The earth's coal supplies are expected to last for two or three centuries, but petroleum will run out in 70 to 80 years. At the present time, about 60 percent of the industrial energy for the world (67 percent for the United States) comes from oil and gas. These figures suggest that many changes are due in the relatively near future.

Natural gas is the cleanest of the fossil fuels, and it will probably be the first to run out. Coal, which is the most widely used, releases sulfur and mercury. Sulfur is a major problem in some areas where it causes atmospheric pollution. Use of low-sulfur coal or removal of sulfur from the smoke is required in some cities. It is estimated that 3000 tons of toxic mercury are also released from burning coal each year. This is about equal to the amount lost in industrial processes, and over ten times the amount released by natural weathering.

Water is an attractive power source because it does not pollute the atmosphere. Damming rivers, however, spoils some of their esthetic value, and the lakes formed by the dams will ultimately fill with silt. Transporting the power from generally remote hydroelectric plants requires transmission lines. In spite of these problems, water power will probably be important in Africa and South America because these continents lack large supplies of coal but have large sources of unused water power. Water power could supply the earth with energy at the present rate of usage, but because of the silting of the lakes, it could only do so for one or two centuries.

Solar radiation as a source of energy is attractive because no fuel is used. The disadvantages are that large areas are required to capture sufficient solar radiation and that only certain areas, such as deserts, receive enough constant sunlight to make such systems feasible. These disadvantages are offset somewhat because desert areas are largely unused for any other purpose, although some ecologists would disagree. They feel that deserts, like mountain, forest, and beach areas, should be preserved for recreational, ecologic, and esthetic reasons.

The relatively cloudless deserts of the southwestern United States receive about 0.8 kilowatt per square meter for the middle six to eight hours of the day most of the year. A black surface can absorb most of this energy, but a black body is also an excellent radiator. Recently, "one-way" surface coatings have been developed that are transparent to the sun's radiation, allowing it to pass through and be absorbed, but are reasonably opaque to the longer-wavelength radiation from the collector. A collector could heat molten salts that would maintain a fairly constant temperature during overnight operation. A large-scale experiment will be necessary to prove the practicality of such a system.

Tidal energy and wind power can supply only a small amount of our energy needs, and no really practical method of harnessing them has been devised. Their advantages are no fuel, no waste, no pollution, and minimal ecologic and scenic damage.

Geothermal energy is a new energy source in the United States, but has been used in Italy since 1904. The number of volcanic areas where geothermal steam is available is small. The life of a steam well is also limited. The best current estimates suggest that about one-fifth of our present energy needs could be met for about 50 years by geothermal power.

Nuclear energy may be the energy source of the future. However, three problems must be overcome: they are disposal of radioactive wastes, development of reactors that use the more plentiful types of ores, and development of better safeguards to prevent human error in operation. Other serious obstacles are public acceptance of nearby nuclear reactors, disposal of unused heat, and assurance that fissionable products will be used for peaceful purposes.

KEY TERMS

Minerals

Element
Atom
Mineral
Crystalline
Crystal
Rock
Ion
SiO_4 tetrahedron
Feldspar
Plagioclase feldspar
Potassium feldspar
Perthite
Olivine
Pyroxene
Augite
Amphibole
Hornblende
Mica
Biotite
Muscovite
Chlorite
Quartz
Color
Streak
Luster
Metallic
Nonmetallic
Submetallic
Fracture
Conchoidal
Cleavage
Hardness
Specific gravity
Igneous rock
Metamorphic rock
Sedimentary rock
Rock cycle

Igneous Rocks

Magma
Reaction series
Differentiation
Continuous series
Discontinuous series
Texture
Intrusive
Extrusive
Concordant
Discordant
Batholith
Dike
Sill
Laccolith
Porphyritic
Pumice
Pyroclast
Ash grain
Lapilli
Block
Bomb
Tephra
Pyroclastic rocks

Tuff
Lapilli tuff
Pyroclastic breccia
Agglomerate
Epiclastic tuff
Epiclastic breccia
Welded tuff
Shield volcano
Composite cone volcano
Pegmatite
Oceanic crust
Mantle
Peridotite
Dunite
Granite
Rhyolite
Felsite
Diorite
Andesite
Obsidian
Gabbro
Basalt

Sedimentary Rocks

Mechanical weathering
Chemical weathering
Frost wedging
Oxidation
Hydrolysis
Carbonation
Leaching
Soil
Soil horizon
Humus
Pedalfer
Pedocal
Caliche
Laterite
Bauxite
Lithification
Cementation
Compaction
Recrystallization
Mud crack
Ripple mark
Cut and fill
Cross-bedding
Graded bed
Sorting
Porosity
Permeability
Clastic
Conglomerate
Breccia
Sandstone
Siltstone
Mudstone
Shale
Clastic rocks
Nonclastic rocks
Limestone
Dolomite
Chert
Rock salt
Evaporite
Chalk
Diatomite
Geosyncline
Continental shelf

Metamorphic Rocks

Pressure
Stress
Shear
Thermal or contact metamorphism
Hornfels
Regional metamorphism
Foliated rocks
Gneiss
Schist
Dynamic metamorphism
Mylonite
Marble
Quartzite
Progressive metamorphism
Retrogressive metamorphism
Serpentinite
Soapstone
Talc schist
Slate
Greenschist
Amphibolite

QUESTIONS

1. Name the four most abundant elements (weight percent) in order. Which has the greatest volume?
2. Define mineral.

3. Distinguish between crystal and crystalline.
4. Why are there only eight important rock-forming minerals?
5. Can you relate the perfect cleavage in mica to its internal structure?
6. How are minerals distinguished from rocks?
7. What happens when a crystal of calcic plagioclase is melted?
8. Describe the crystallization of plagioclase in a magma.
9. Name an ore of zinc, of iron, of lead.
10. Distinguish between silicon and silicate.
11. What factors determine whether one element can substitute for another element in a mineral?
12. Define igneous rock.
13. Describe the crystallization of a basaltic melt.
14. How can a sill be distinguished from a flow?
15. What cooling history does a porphyritic texture imply?
16. What is a batholith? What is the usual composition of a batholith, and what problems do batholiths present?
17. Name a concordant igneous body.
18. In what type of occurrence (sill, flow, batholith, etc.) would you expect to find glassy, fine-grained, porphyritic, and coarse-grained igneous rocks?
19. Describe the formation of pumice.
20. Describe the types of volcanoes. Include how they are formed and which is steepest and why.
21. What is the probable origin of the mid-ocean ridge basalt?
22. What is the probable origin of andesite at volcanic island arcs?
23. What minerals are most likely to be visible in a felsite porphyry?
24. Describe the formation of ore veins.
25. Igneous rocks are classified on the bases of ______________ and ______________. Is this a genetic classification? Discuss.
26. Clastic sedimentary rocks are classified on the basis of ______________.
27. In what climate is chemical weathering most effective? Why?
28. In what climate is mechanical weathering most effective? Why?
29. Describe two methods of mechanical weathering; of chemical weathering.
30. List the common minerals in the order of their stability to chemical weathering. Why is quartz a common mineral in sandstone?
31. What is the origin of chert?
32. What are the common cements in sandstones?
33. How can the top and bottom of beds be determined in a group of vertical sedimentary beds?
34. Define metamorphic rock.
35. What is foliation, and how does it form?
36. What is contact metamorphism, and what types of rocks are formed?
37. How does a shale differ from a slate?
38. What is the low-grade metamorphic equivalent of a basalt?
39. Serpentinite forms from metamorphism of ______________.
40. Soapstone forms from metamorphism of ______________.
41. Name the following rocks and interpret their significance as completely as possible:
    a. Coarse-grained rock with interlocking texture composed mainly of quartz and feldspar with some dark minerals.

b. Fine-grained, dark-gray rock with a few well-formed plagioclase and olivine crystals.
c. Foliated rock with much muscovite.
d. Fine-grained, white rock composed of calcite with some clam shells.
e. Clastic rock composed of well-rounded quartz grains about 1 mm in diameter.
f. Clastic rock composed of quartz and feldspar with some hornblende and mica. The grains are mainly between 1 and 2 mm in diameter and are angular. (*Note:* After reading the next chapter, you may want to review your interpretations of the last two rocks.)

42. Before starting to identify rock specimens, it may help to list all of the rocks mentioned in Chapter 1 in the form suggested below. This will show the importance of *texture* in recognition and interpretation of rocks.

Check which type: Ig Sed Meta

Coarse-grained

Clastic texture

____

____

Interlocking texture

Massive ________________________________

____

Foliated ________________________________

____

Fine-grained

____

____

Glassy ________________________________

Frothy ________________________________

43. Rockhounds collect many different rocks and minerals that they cut and polish. What physical properties should such rocks and minerals have? List the rocks and minerals that have such properties.
44. Discuss whether the earth's mineral resources are sufficient for the next hundred years.
45. List the earth's energy sources, and discuss the future of each.

SUPPLEMENTARY READING

General

Dietrich, R. V., and B. J. Skinner. *Rocks and Rock Minerals.* Somerset, NJ: John Wiley & Sons, 1979, 319 pp.

Minerals

Buseck, P. R. "Electron Microscopy of Minerals." *American Scientist* 71(2) (March–April 1983): 175–85.

Klein, Connelis, and C. S. Hurlbut, Jr. *Manual of Mineralogy*. 20th ed. New York: John Wiley & Sons, 1985, 596 pp.

Igneous Rocks

Decker, Robert, and Barbara Decker. "The Eruptions of Mount St. Helens." *Scientific American* 244(3) (March 1981): 68–80.

Decker, Robert, and Barbara Decker. *Volcanoes*. San Francisco: W. H. Freeman & Co., 1981, 244 pp.

Frances, Peter. "Giant Volcanic Calderas." *Scientific American* 248(6) (June 1983): 60–70.

Frances, Peter, and Stephen Self. "The Eruption of Krakatau." *Scientific American* 249(5) (November 1983): 172–87.

Miller, C. D. *Potential Hazards from Future Eruptions in the Vicinity of Mount Shasta Volcano, Northern California*. U.S. Geological Survey Bulletin 1503. Washington: U.S. Government Printing Office, 1980, 43 pp.

Newhall, C. G., and Daniel Dzurisin. *Historical Unrest at Large Calderas of the World*. 2 vols. U.S. Geological Survey Bulletin 1855. Washington: U.S. Government Printing Office, 1980, 1108 pp.

Sheets, P. D., and D. K. Grayson, eds. *Volcanic Activity and Human Ecology*. New York: Academic Press, 1979, 644 pp.

Sedimentary Rocks

Land, L. S. "The Origin of Massive Dolomite." *Journal of Geological Education* 33(2) (March 1985): 112–25.

Millot, Georges. "Clay." *Scientific American* 240(4) (April 1979): 108–18.

Pettijohn, F. J., and P. E. Potter. *Atlas and Glossary of Primary Sedimentary Structures*. New York: Springer-Verlag, 1964, 370 pp.

Reading, H. G., ed. *Sedimentary Environments and Facies*. New York: Elsevier, 1978, 557 pp.

Weathering

Lal Gauri, K. "The Preservation of Stone." *Scientific American* 238(6) (June 1978): 126–36.

Keller, W. D. "Geochemical Weathering of Rocks: Source of Raw Materials for Good Living." *Journal of Geological Education* 14(1) (February 1966): 17–22.

Severson, R. C., and H. T. Shacklette. *Essential Elements and Soil Amendments for Plants: Sources and Use for Agriculture*. U.S. Geological Survey Circular 1017. Washington: U.S. Government Printing Office, 1988, 48 pp.

Metamorphic Rocks

Dietrich, R. V. "Migmatites—A Résumé." *Journal of Geological Education* 22(4) (September 1974): 144–56.

Higgins, M. W. *Cataclastic Rocks*. U.S. Geological Survey Professional Paper 687. Washington: U.S. Government Printing Office, 1971, 97 pp.

Resources

Brobst, D. A., and W. P. Pratt, eds. *United States Mineral Resources.* U.S. Geological Survey Professional Paper 820. Washington: U.S. Government Printing Office, 1981, 722 pp.

Eaton, G. P. "Mineral Abundance in the North American Cordillera." *American Scientist* 72(4) (July–August 1984): 368–77.

Haymon, R. M., and K. C. MacDonald. "The Geology of Deep-Sea Hot Springs." *American Scientist* 72(5) (September–October 1985): 441–49.

Heiken, Grant, et al. "Hot Dry Rock Geothermal Energy." *American Scientist* 69(4) (July–August 1981): 400–07.

Hekinian, R., et al. "Sulfide Deposits from the East Pacific Rise Near 21°N." *Science* 207(4438) (March 28, 1980): 1433–44.

Menard, H. W. "Toward a Rational Strategy for Oil Exploration." *Scientific American* 244(1) (January 1981): 55–65.

Meyer, Charles. "Ore Metals through Geologic History." *Science* 227(4693) (March 22, 1985): 1421–27.

U.S. Geological Survey. *National Energy Resource Issues: Geologic Perspective and the Role of Geologic Information.* U.S. Geological Survey Bulletin 1850. Washington: U.S. Government Printing Office, 1988, 79 pp.

2

The External Processes—Erosion

The outer few kilometers of the earth are called the crust. (We will define the term later.) Although the crust is the only part of the earth accessible to direct observation, studying the earth by observations in the crust is like trying to read a book by its covers. Actually, the problem is even greater because 71 percent of the surface is covered by oceans.

Classical geology developed from observations made on the continents. One of the most fundamental ideas or laws to have been developed is that of **uniformitarianism,** which says that *the present is the key to the past.* In other words, any structure in old rocks must have been formed by processes similar to those now going on upon the earth. Although this simple statement of uniformitarianism has limitations, it was very important to the development of geology. Uniformitarianism has not always been accepted. We see erosion going on around us, note the material carried by muddy rivers, and conclude from this that rivers eroded the valleys in which they flow. However, not too many years ago, some geologists believed that valleys—especially deep valleys such as Yosemite in California—were formed by great earthquakes which split open the earth, and that the rivers then flowed in them simply because they were the lower areas.

Erosion is clearly the main process occurring today on the earth's surface, and it is shown by gullies on hillsides, landslides, and similar examples. The agents of erosion, of which rivers are by far the most important, are constantly wearing down the continents. Nevertheless, we still have mountains. The present rate of erosion is such that, theoretically, all topography will be removed in about 12 million years. However, we know that erosion and deposition have been going on for a few billion years, so something must interrupt the work of the rivers and uplift the continents. Thus, much of geology is concerned with the struggle between the external forces of erosion and the internal forces that cause uplift.

Some familiar demonstrations of the internal forces causing uplift are earthquakes, volcanoes, and folded and tilted sedimentary rocks that are now high on mountains but were originally formed as flat beds at or below sea level. Physical or dynamic geology is the study of this battle between the internal forces of uplift and the external forces of erosion. We will consider these two forces separately in order to eliminate confusion, but remember that they go hand in hand.

The underlying cause of erosion is downslope movement under the influence of gravity, and the agents that cause such movements are water (rivers, etc.), ice (glaciers), and wind. The material that is moved by these agents comes from the weathering processes discussed earlier.

DOWNSLOPE MOVEMENT OF SURFACE MATERIAL

All of the surface material of the earth tends to move downslope under the influence of gravity. *The surface material is more-or-less weathered and is generally termed* **overburden** to differentiate it from *bedrock.* These downslope movements shape the earth's surface and so must be understood in order to interpret landscapes. Downslope movement generally ends at a river or stream, which carries away the material. Thus, rivers and downslope movements work together to shape the landscape. A small example of this, discussed in the section on rivers, shows that, although a river downcuts to form a valley, the shape of the valley sides is determined by downslope movements, which widen the narrow cut made by the river.

The efficiency of downslope movements is easily shown by analysis of slopes. Such analysis shows that most of the earth's surface slopes less than 5 degrees and that very little of the earth's surface slopes more than 45 degrees. Many forms of downslope movement are slow processes. However, when slopes are disturbed by buildings, roads, or irrigation, more rapid movements may be triggered, causing much damage. Thus, knowledge of these processes is important when any type of engineering project is considered on even moderately sloping ground (Fig. 2.1).

Downslope movements may be rapid or extremely slow; they may involve only the very topmost surface material, or they may be massive movements that

FIGURE 2.1
Hillside construction can cause earth movements if not properly planned.

FIGURE 2.2
The slope formed by debris at the foot of a cliff is called a talus. Mechanical weathering generally produces the debris.

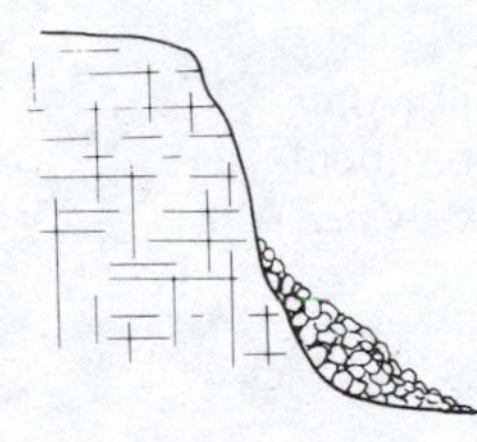

involve the total thickness of overburden or even bedrock. The processes at work on any slope depend on many factors. *On steep slopes no overburden generally can develop; therefore, the dominant process on such slopes is* **rockfall,** *and the slope that develops at the foot is called a* **talus** (Fig. 2.2). On less steep slopes (how much less depends on the climate, which controls weathering processes), overburden may develop, and this material may move downslope in a number of ways discussed below. Thus, weathering determines the type of downslope movement, thereby causing the different aspects of topography in a humid area and a dry area (Fig. 2.3). The structure and rock types on a hill also may control erosion processes. Differences in rate or kind of weathering can determine the amount of material available for downslope movement, and so the underlying bedrock structure may be etched in relief.

Falling rain erodes surface material in two ways. The impact of raindrops loosens material and splashes it into the air, and on hillsides this material falls back at a point lower down the hill (Fig. 2.4). The water that does not infiltrate runs down the hillside and erodes in the same manner as the water in a river. The amount of erosion caused in these ways depends on the steepness of the slope and on the type and amount of vegetative cover. The effects of such erosion are best seen in areas where the vegetation has been removed by fires or people.

The massive types of downslope movement that involve appreciable thicknesses of material are creep, earthflow, mudflow, and landslide.

Creep *is the slow downslope movement of overburden* and, in some cases, bedrock. It is recognized by tilted poles and fence posts if the movement is great

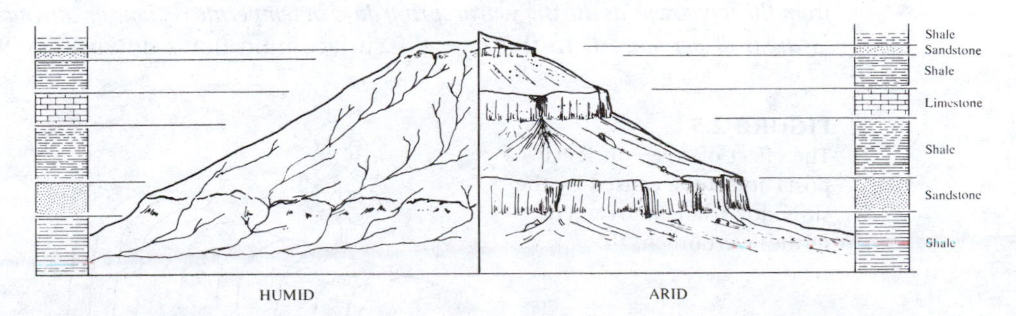

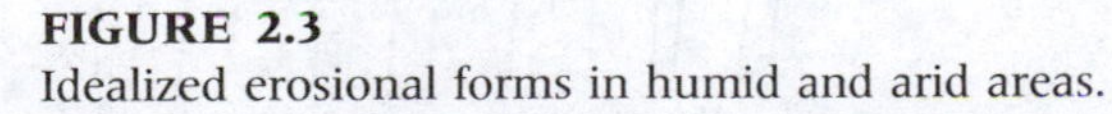
FIGURE 2.3
Idealized erosional forms in humid and arid areas.

FIGURE 2.4
Erosion by raindrop. (After U.S. Department of Agriculture 1955 yearbook, *Water.*)

enough; commonly in such cases, trees will be unable to root themselves, and only shrubs and grass will grow on the slope. In other cases, where the creep is slower, the trunks of trees will be bent as a result of the slow movement (Fig. 2.5).

If the amount of water in the overburden increases, an **earthflow** *results. In this case a tongue of the overburden breaks away and flows a short distance.* An earthflow differs from creep in that a distinct, curved scarp is formed at the breakaway point (Fig. 2.6).

With increasing water, an earthflow may grade into a **mudflow.** *The behavior of mudflows is similar to that of fluids.* Rain falling on the loose pyroclastic material on the sides of some types of volcanoes produces mudflows, and this is a very important mode of transportation of volcanic material. Another type of mudflow is common in arid areas. In this case, a heavy thunderstorm produces large amounts of runoff in the drainage area of a stream. The runoff is in the form of rapidly moving sheets of water that pick up much of the loose surface material. Because these sheet floods flow into the main stream, all of this muddy material is concentrated in the main stream course. As a result, a dry stream bed is transformed into a flood very rapidly. This flood of muddy material is a mudflow and moves very swiftly, in some cases with a steep, wall-like front. Such a mudflow can cause much damage as it flows out of the mountains. Eventually, loss of water (generally by percolation into the ground) thickens the mudflow to the point that it can no longer flow.

Another type of mudflow called **solifluction** *occurs when frozen ground melts from the top down, as during warm spring days in temperate regions or during the summer in areas of permafrost.* In this case the surface mud flows downslope. Solifluction

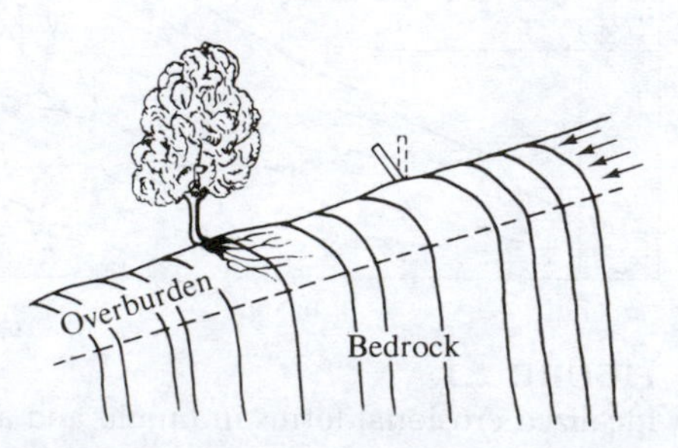

FIGURE 2.5
The effect of creep on fence posts and trees, caused by the slow downslope movement of the overburden.

FIGURE 2.6
An earthflow, showing the curved scarp at the break-away point.

causes many problems in construction, especially in the far north where permafrost occurs. Permafrost areas, as the name implies, are areas where the deeply frozen ground does not completely melt during the summer.

Landslides *are rapid slides of bedrock.* They are of two types: the *landslide* or *rockslide,* and the *slump.* **Rockslides** *develop when a mass of bedrock breaks loose and slides down the slope* (Fig. 2.7). In many cases the bedrock is broken into many fragments during the fall, and this debris behaves as a fluid and spreads out in the valley. It may even flow some distance uphill on the opposite side of the valley, if the valley into which it falls is narrow. Such landslides are sometimes called *avalanches,* but this term is best reserved for snowslides. Landslides generally are large and destructive, involving millions of tons of rock. As suggested in the diagrams, landslides are apt to develop if planes of weakness, such as bedding or jointing, are parallel to a slope, especially if a river, a glacier, or people undercut the slope. Thus, landslide danger can be evaluated by geologic study.

Slumps *tend to develop in cases where a strong, resistant rock overlies weak rocks.* Note in Figure 2.8 the curved plane of slippage, the reverse tilt of the resistant unit that may provide a basin for a pond to develop, and the rise of the toe of the slump which is most pronounced where the underlying rock is very weak and can flow plastically. Unlike rockfalls, slumps develop new cliffs nearly as high as those previous to the slump, thus setting the stage for a new slump. Therefore, slumping

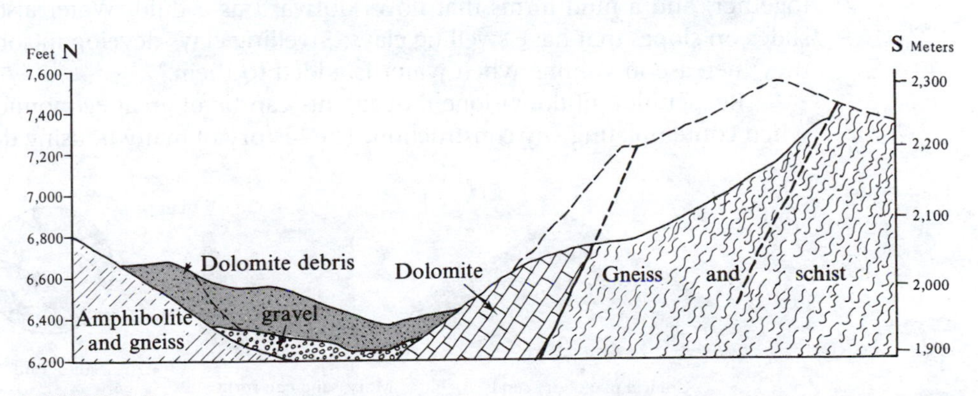

FIGURE 2.7
Cross section of a landslide into the Madison River valley near Hebgen Lake, Montana. The slide was triggered by an earthquake and came from the right, leaving a large scar on the mountain side. The debris traveled uphill on the left side of the valley. (After J. B. Hadley, U.S. Geological Survey Professional Paper 435, 1964.)

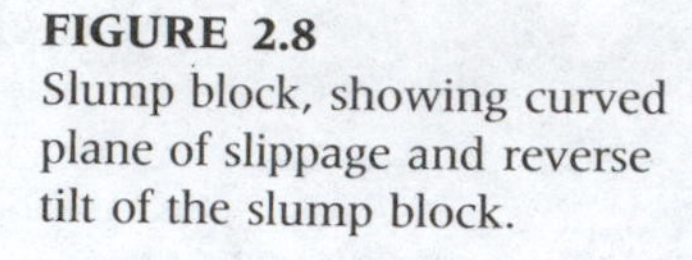

FIGURE 2.8
Slump block, showing curved plane of slippage and reverse tilt of the slump block.

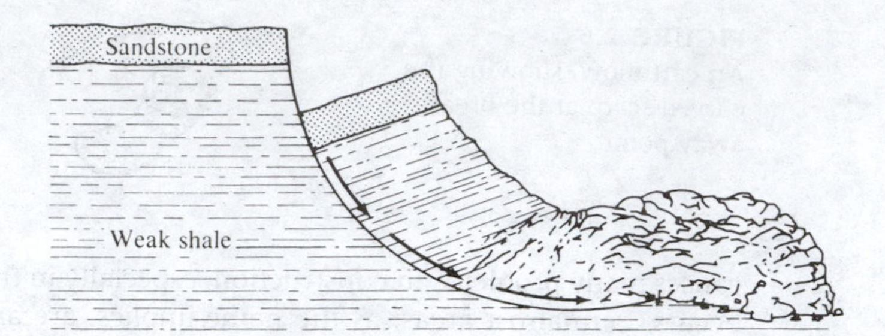

is a continuous process, and generally in areas of slumping, many generations can be seen far in front of the present cliffs.

Massive (downslope) movements may be triggered in many ways. The most common are (a) undercutting of the slope, (b) overloading of the slope so that it cannot support its new weight and, hence, must flow or slide, (c) vibrations from earthquakes or explosions that break the bond holding the slope in place, and (d) additional water. The addition of water has several effects and is the reason for many newly made cuts standing only until the following spring. The effect of water is twofold; it adds to the weight of the slope, and it lessens the internal cohesion of the overburden. Although water's effect as a lubricant is commonly considered to be its main role, this effect actually is slight. The main effect is lessening the cohesion of the material by filling the spaces between the grains with water.

A simple experiment with sand illustrates this point. Dry sand can be piled only in cones with slopes slightly over 30 degrees. Damp sand, however, can stand nearly vertically because the small amounts of moisture between the grains tend to hold the grains together by surface tension (Fig. 2.9). When enough water is added to completely fill the intergrain voids, no surface tension exists to hold the grains together, and a mud forms that flows outward as a fluid. Water also can trigger slides on slopes that have swelling clays. Swelling clays develop in some soils, and they increase in volume when water is added to them.

Recognition of downslope movements can be of great economic importance when contemplating any construction. The history of many housing developments

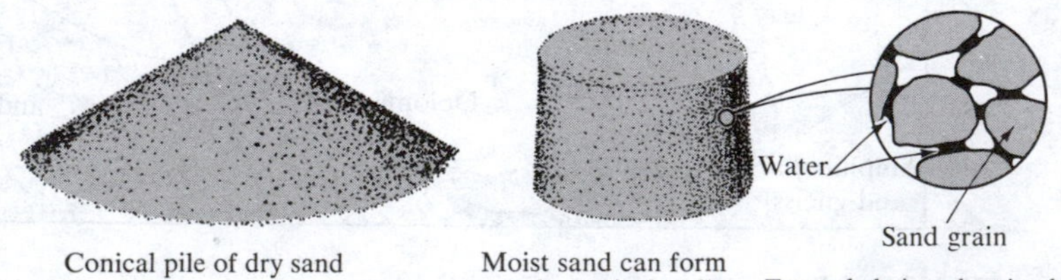

FIGURE 2.9
The effect of water on the cohesion of sand.

suggests that even house or lot buyers could benefit from the ability to recognize them. Generally, downslope movements are best detected by observation from a distance, followed by detailed close study. A slope in motion generally will have few trees and may have a bumpy, hummocky appearance. Because some types of glacial deposits also have this appearance, one should know how to recognize glacial topography, which will be discussed later.

Landslides can be stabilized in a number of ways. The most important point to make is that all of these methods are much more economical during the initial construction rather than as a first-aid measure later on. In many cases it is possible, with very minor redesign of any engineering project, to avoid potential landslide problems. One very common misconception is that landslides can be stabilized with planting. Indeed, the author has personally met a number of homeowners in landslide areas who felt that growing either grass or similar ground covers would stabilize a landslide area. This is a complete misconception, and there are numerous examples of very large trees (even forests) moving as parts of landslide blocks. Many landslides move at depths greater than the normal root depth of most trees.

The main methods of stabilizing landslides are excavation, drainage, construction of retaining walls or similar structures, use of rock bolts, and injection of grout into the slide. Bridging of highways over unstable areas can avoid landslide damage. Most common of all these methods is to drain the landslide. Generally this is done by drilling holes into the slide and providing perforated pipes through which the water can escape. In some cases this is coupled with an impervious layer over part of the landslide to prevent the entrance of surface water into the slide. In a few cases the slides are deliberately triggered by blasting to stabilize the area.

THE GEOLOGIC WORK OF RIVERS

Rivers are agents of erosion, transportation, and deposition. That is, they carve their own valleys and carry the eroded material downstream, where it is either deposited by the river or delivered into a lake or the ocean. Rivers are the most important agents in transporting the products of erosion. They drain vastly larger areas than do glaciers, and, even though they flow infrequently in deserts, they are able to carry more material than can the ever-present wind. Thus, it is rivers that receive and transport the products of weathering that are fed into them by runoff from precipitation as well as by downslope movement. Weathering, downslope movement, and rivers work together to shape the landscape.

Graded Rivers

Most rivers are in balance and are called **graded rivers.** The balance is shown in a number of ways. One obvious illustration is that all the tributary streams meet the main stream at the level of the main stream. This accordance of junctions is too universal to be due to chance. (This accordance is not true in glaciated areas, however, where stream levels have been changed by glacial erosion.) A graded

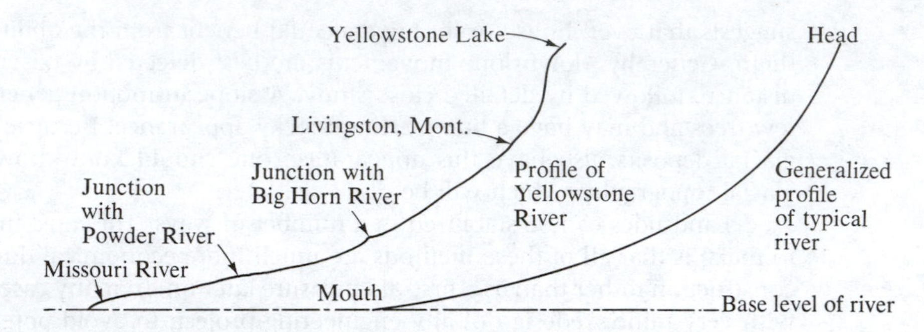

FIGURE 2.10
Longitudinal river profiles. Vertical scale exaggerated. (Yellowstone River profile after U.S. Geological Survey Water Supply Paper 41, 1901.)

stream has just the right gradient to be able to transport the debris delivered to it by its own erosion, its tributaries, downslope movement, and rain wash. This ability is a long-term average, so that a given stretch of river at one season may be eroding its bed, and at another time may be building up its bed. These changes in regime vary from year to year with such climatic influences as amount and intensity of precipitation.

River surveys show that the profile from head to mouth of many streams is a smooth curve that gradually steepens toward the head (Fig. 2.10). We may rationalize this profile by remembering that at its head a stream has small flow and steep gradient to give it eroding and transporting ability, whereas downstream it has larger flow and lower gradient. The profile is most concave on rivers where abrasion reduces the average size of transported particles as they move downstream.

In most river systems the profile is more complicated because the ability of a river to erode and transport depends largely on the velocity, the amount of flow, the gradient, and the shape and roughness of the channel. At the point where a tributary joins the main stream, the regime of the main stream is abruptly changed by the additional flow and the increased debris from the tributary. The main stream reacts to this change by changing its gradient so that it can handle the new material. These changes to the simple profile are shown on the profile of the Yellowstone River in Figure 2.10.

The limit to which the river can erode is, of course, the elevation of its mouth. This limit is called the **base level** *of the river*. For most big rivers the base level is sea level, but for some that flow into lakes, it is the elevation of the lake. For tributary streams the base level is the elevation at which they join the main streams.

Erosion and Transportation by Rivers

Rivers erode by several processes, depending on the nature of the bedrock. If the bedrock, or parts of it, is soluble, **solution** will occur. Hard, resistant bedrock is

FIGURE 2.11
Potholes in a river bed cut in granitic rock. The potholes are formed during times of much greater flow. (Photo by Mary Hill, Courtesy California Division of Mines and Geology)

mechanically eroded by **abrasion** by particles moved by the water. A common type of abrasion results in cylindrical holes, called **potholes,** worn into the bedrock, apparently by stones that are spun by eddy currents (Fig. 2.11). Softer bedrock and loose material are moved by the pressure and shearing effect of the moving water.

The material carried by a river can be considered in three parts—*dissolved load, suspended load, and bed load.* The **dissolved load** *is soluble material,* and its presence has no effect on the regime of the stream. The **suspended load** *is the fine material that, once in the water, settles so slowly that it is carried long distances.* As the velocity of a river increases, some of the finer bed load becomes suspended load so that the distinction can be made only at a given moment. Suspended material is the cause of muddy river water. Clay particles settle very slowly and so are carried relatively long distances. Silt settles much more rapidly; it is carried as suspended load only by turbulent water, which has enough upward currents to keep the silt from settling. Thus, clay is carried farther than silt, resulting in sorting of sediment by rivers. The bed load is not only sorted; its particles also are worn during transportation. The **bed load** *is moved by sliding and rolling along the bottom.* In the summer or fall, it is difficult to believe that the large boulders in a stream bed, particularly of a small stream, are part of the bed load and are being moved by the stream. We must remember that the largest boulders may move only during periods of very great flow, which may occur only a few times each century. However, if we visit such a stream during the spring when it is filled with meltwater, we commonly can hear "clicks" when a large rock rolls or slides into its neighbor.

At the headwaters, rivers actively erode, mainly by downcutting. This downcutting lengthens the stream (Fig. 2.12). At places, this process produces interesting results when an actively headward-eroding tributary to one river system meets a trunk stream of another river system. When this happens, *the steeper gradient of the eroding stream captures the flow from the trunk stream. This process is called* **stream piracy** (Fig. 2.13).

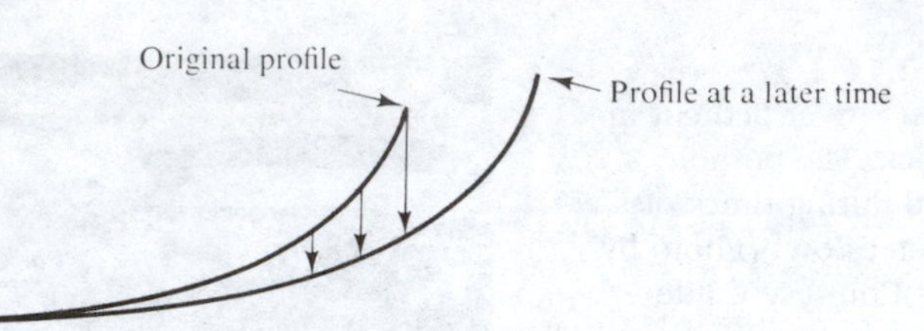

FIGURE 2.12
Downcutting at headwaters of a stream lengthens the course of the stream.

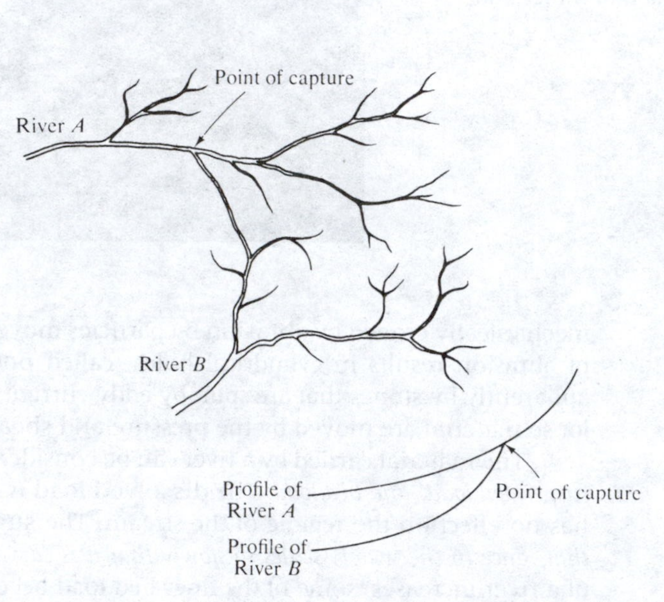

FIGURE 2.13
A common type of stream piracy. Rapidly downcutting minor tributary of River *B* captures the headwaters of River *A*. The gradients of the two rivers near the point of capture are shown in the lower diagram.

River valleys are widened largely by *lateral cutting* on the outside of curves, and the valley sides are shaped by downslope movement of material. Thus, curves which, as shown in Figure 2.14, result from the many irregularities on a river, broaden the river valley. Also as shown in Figure 2.14, there is a strong tendency to develop *regular, repeating curves called* **meanders,** especially if the valley floor is easily erodible, as is generally the case in a broad valley.

Meanders are but one type of river course. Study of most rivers will show changes from curves to straight reaches or meanders from place to place. At other places *the same river may divide into many channels separated by bars; such places are termed* **braided** (Fig. 2.15). Another feature, familiar to those who fish, is the alternation of riffles and pools on most rivers. Recent studies have suggested that all of these features develop in such a way that the energy of the river is used uniformly.

River Deposition

Rivers deliver most of their loads to lakes or to the ocean. River deposits in the form of **deltas** develop where rivers enter standing water; even in this case, however,

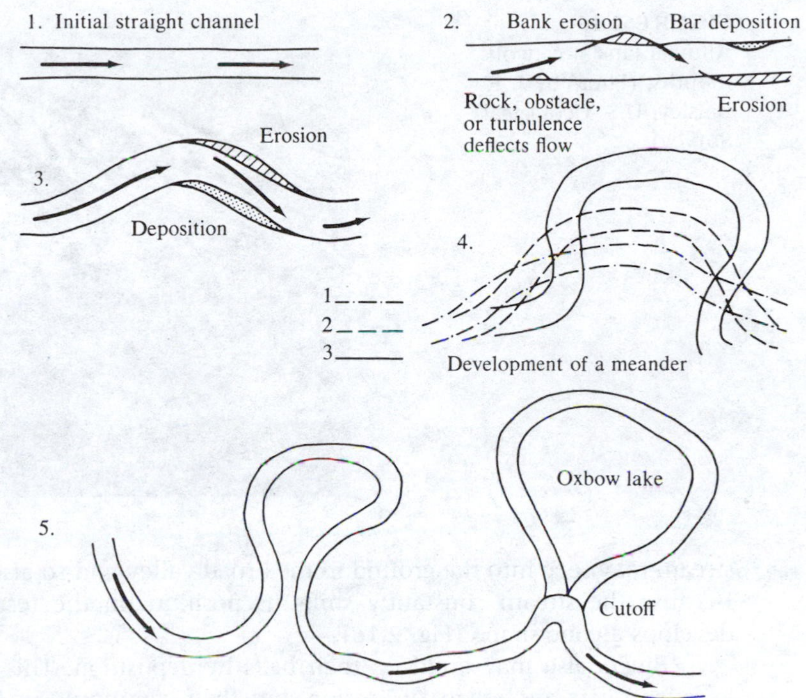

FIGURE 2.14
Development of meanders.

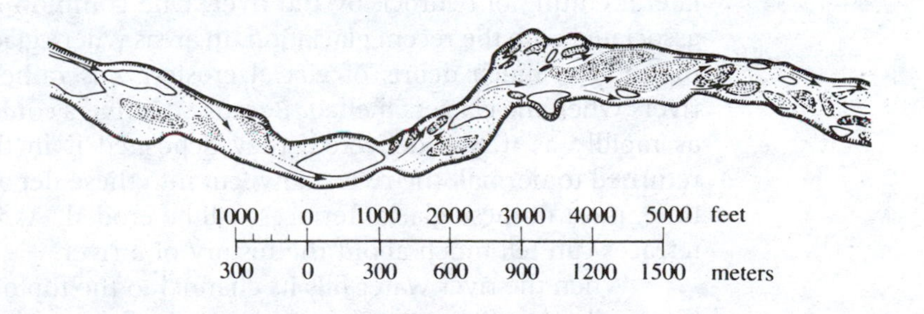

FIGURE 2.15
A braided part of Middle Loup River in Nebraska. Sandbars are stippled and islands are blank. Arrows show the flow. (After J. C. Brice, U.S. Geological Survey Professional Paper 422-D, 1964)

much of the river's load may be reworked by offshore currents, thus becoming marine rocks or lacustrine (lake-deposited) rocks. Deltas form because the velocity of river water is slowed so abruptly on entering standing water that the bed load is dropped. The dissolved and suspended loads may be carried farther out.

Alluvial fans *are similar to deltas and are formed where streams flow out of mountains into broad, relatively flat valleys.* In this case, the flatness of the valley slows the stream, causing deposition. In addition, especially in arid areas, the mountain

FIGURE 2.16
Alluvial fans at canyon mouths. (Photo by J. R. Balsley, U.S. Geological Survey)

stream may seep into the ground in the broad valley and so also cause deposition. Because the stream constantly shifts its position on the resulting fan, the fan develops a cone shape (Fig. 2.16).

Rivers also may build up their beds by deposition. This may seem strange because rivers are agents of erosion, but they commonly do so for a variety of reasons. *In some cases a river may deposit during a long period of time, so that its meandering causes deposition that covers the entire valley. Such a process produces a smooth surface. Later, due to some change in its regime, the river may cut into this surface and so form one or more lower* **terraces** (Fig. 2.20). Terraces also can be formed by lateral cutting of bedrock by the river. One common cause of terrace building is associated with the recent glaciation. In areas where glaciers used to be active in the mountains, much debris of glacial erosion was rather suddenly released to the rivers when the glaciers melted. Because the rivers could not transport the material as rapidly as it was delivered, they deposited it in their valleys. As conditions returned to normal, the rivers downcut into these deposits and formed terraces. In time, most of these glacial terraces will be eroded. As is shown later, the study of terraces can tell much about the history of a river.

When the river water fills its channel to the top of the banks, any increase in flow will cause flooding. The floodwater spills over the river banks, filling part or all of the valley. The river is doing its maximum erosion and transportation at the time of flood, and as the floodwater spreads out, its velocity is quickly reduced, causing deposition near the banks of the river. Such deposits are termed **levees** (or **natural levees,** to distinguish them from levees constructed for flood protection)—Figure 2.17. Although the levees help keep the river within its banks, the area between the levee and the side of the river valley may be lower than the river at flood stage; this area is subject to inundation when the flooding river crests above the levee top. Because the river valleys are prime agricultural areas, and because

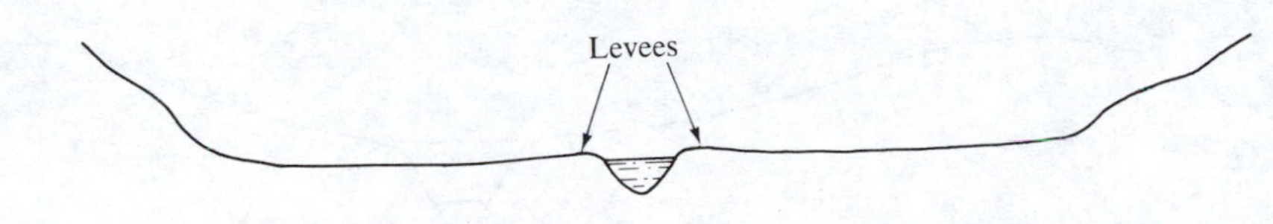

FIGURE 2.17
River levees. Vertical scale exaggerated.

the mechanism of levee-building provides a continuing supply of new soil, many people accept periodic flooding and choose to live in the valleys of large rivers.

River Valleys

The development of a river valley is continuous, involving all of the processes described. As it develops, a valley changes from narrow to broad, as shown in Figure 2.18.

A natural river will not have the same shape throughout its entire length. Areas of soft bedrock will develop faster than areas of resistant bedrock. Because of the increased flow, stretches below major tributaries tend to develop faster than stretches above, so that development generally is faster downstream. Conversely, the headwaters of a stream remain in youth longer.

Rivers can be used to interpret geologic history. Evidence of uplift can be found in individual river valleys because river valleys widen as erosion progresses. Thus, if a river is cutting a V-shaped, youthful valley within a much wider valley,

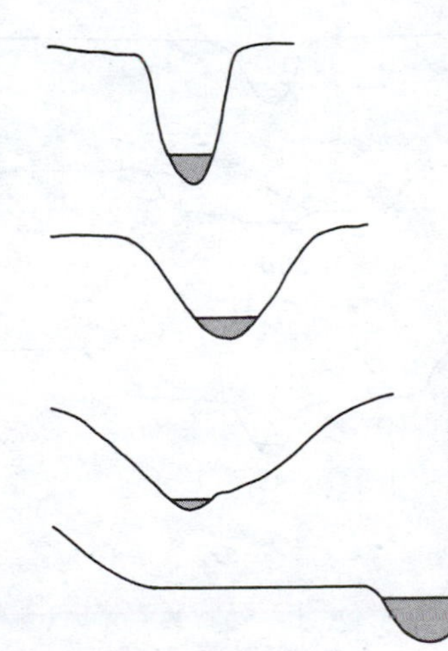

FIGURE 2.18
The development of river valleys. The terms youth, maturity, and old age are sometimes used to describe valleys.

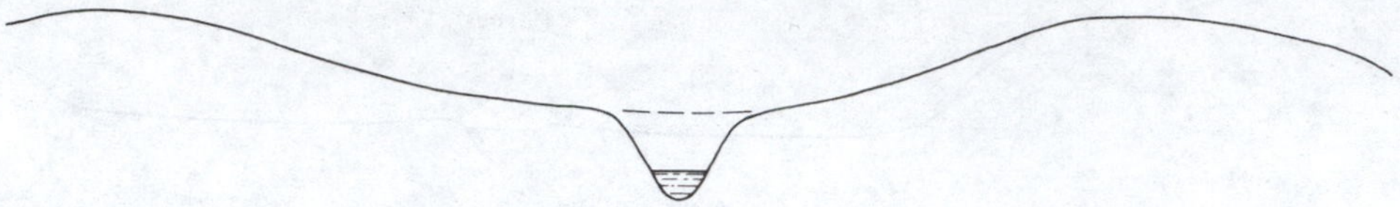

FIGURE 2.19
A broad river valley, with the river cutting a youthful valley within the broad valley. This is generally caused by uplift, but other causes are mentioned in the text.

a change has occurred, causing the river to downcut actively. This change could be due to an uplift, a climatic change, or an increase in the flow of the river from a stream capture or other cause (Fig. 2.19). If no further changes occur, the new valley will widen as erosion progresses. If a new uplift occurs, a new V-shaped valley will be formed. A series of such uplifts is one way of producing river terraces, as these features are called (Fig. 2.20). If depression rather than uplift had occurred, the river valley would have been the site of deposition instead of erosion.

Development of the drainage pattern for the whole drainage basin is an important part of erosion. Development of tributary streams is controlled by many factors, such as type of bedrock, amount of runoff, climate, and vegetation. In many areas of uniform bedrock, branching streams produce a pattern similar to the veins in a leaf (Fig. 2.21). In other areas, the bedrock structure or the jointing may control stream development, and rectangular patterns may develop (Fig. 2.22).

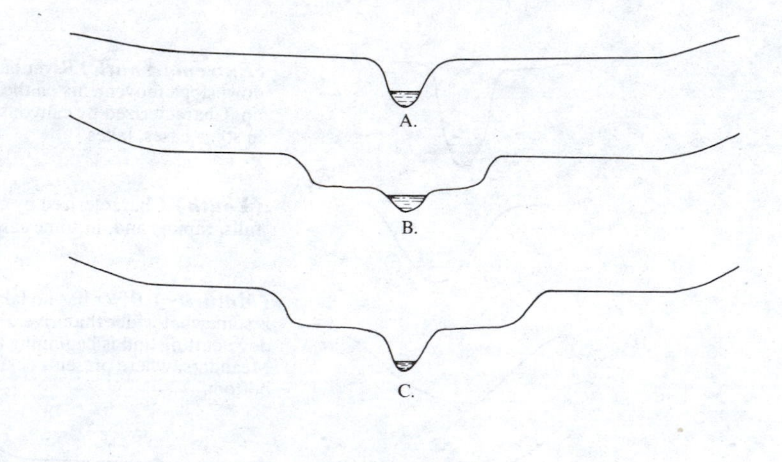

FIGURE 2.20
Successive uplifts have formed a terraced valley. A. An uplift caused the river, initially in a broad valley, to downcut, forming a V-shaped valley. B. Further erosion widened the new valley. C. A new uplift caused renewed downcutting. Ultimately erosion will destroy the older terraces, and it will not be possible to decipher this history.

FIGURE 2.21
Branching stream pattern developed in massive rocks.

FIGURE 2.22
Alternating hard and soft rocks produce a rectangular drainage pattern.

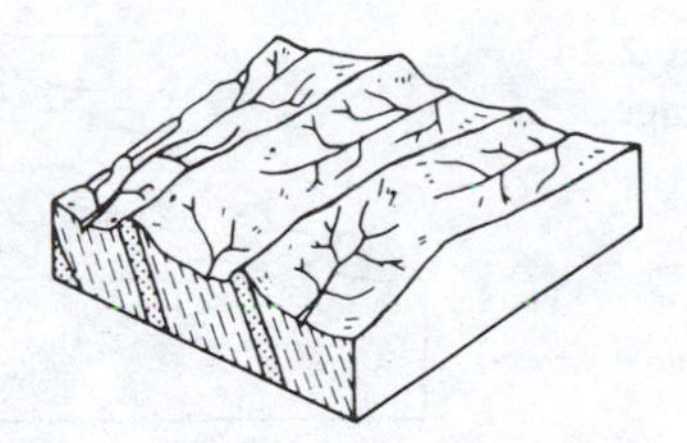

Development of Landscape

The simplest viewpoint on the development of landscape was an American idea, and it had a pronounced influence on the history of geology. The cornerstone of modern geology is the doctrine of *uniformitarianism,* which, in the present context, states that the present processes at work on the earth, given enough time, caused the present landforms. This principle, although recognized earlier by a few workers, was not widely adopted until near the end of the eighteenth century. Geology, however, developed mainly in western Europe during the nineteenth century, and most geologists—especially the island-dwelling English—thought that ocean waves rather than rivers were responsible for most erosion. The opening of the American West after the Civil War provided a vast new area for geologic study. The great canyons and rivers, such as the Colorado, and the many mountain ranges at various stages of dissection, together with a dry climate that has kept bare great vistas, made simple the recognition of huge amounts of erosion (Fig. 2.23).

FIGURE 2.23
A typical scene in the arid southwest. The lack of vegetation and the fact that the flat-lying sedimentary rocks can be traced easily for many miles make recognition of the immense amount of erosion that has occurred easy.

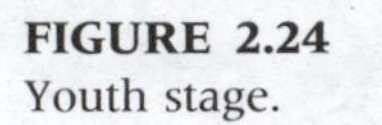
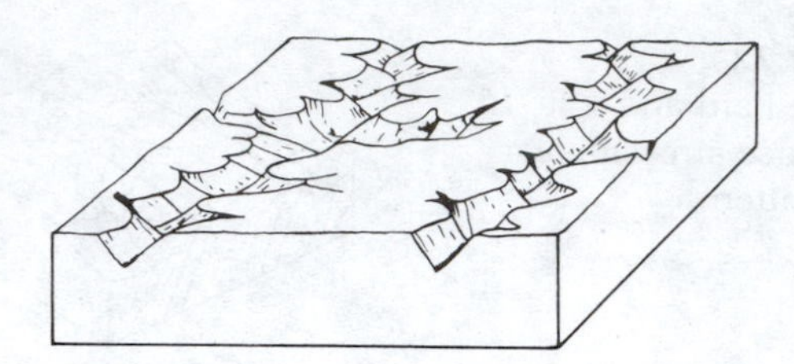

FIGURE 2.24
Youth stage.

Late in the nineteenth century, William Morris Davis of Harvard codified earlier studies into what he termed the *cycle of erosion.* Davis visualized every landscape going through a series of changes, from initial uplift to complete leveling by weathering and erosion. He clearly recognized that an area rarely goes through the complete cycle, but rather that the cycle generally is restarted by new uplifts. This accounts for the fact that there are very few examples of areas that have gone through the complete erosion cycle.

In the **youthful stage,** *a newly uplifted area is relatively flat with deeply entrenched streams* (Fig. 2.24). As erosion proceeds, the interstream areas are reduced in size as a result of downcutting by the rivers and widening of the valleys by downslope movement of material. *When no flat interstream areas exist, the stage has progressed to* **maturity** (Fig. 2.25). This is the stage of maximum relief. *As the interstream areas are rounded and lowered, the area passes into* **old age** (Fig. 2.26). *In the ultimate stage, when even these gentle interstream areas are worn almost flat, the area is called a* **peneplain** (almost a plain, sometimes spelled peneplane). Flat, eroded areas such as peneplains are rare features on the earth's surface, and this fact constitutes one of the weaknesses of the cycle-of-erosion hypothesis. However, many events, such as renewed uplift, can occur to prevent the complete development of a peneplain.

Perhaps the main objection to the Davis cycle is that it requires rapid uplift and then a stillstand during which erosion occurs. If the uplift of an area is slow enough, perhaps erosion keeps pace with the uplift and no change in form occurs. However, any type of uplift from rapid to slow, or from accelerating to slowing, is geologically possible, so, at best, the Davis cycle describes only a special case.

Another objection to the Davis cycle is that the hills might not be reduced in the manner he envisioned. Instead, a balance might result when the hillslopes and stream gradients become adjusted. Such an equilibrium can, in a way, be thought of as an extension of the graded-stream concept discussed earlier. After this stage is reached, the shape of the area remains substantially the same as erosion

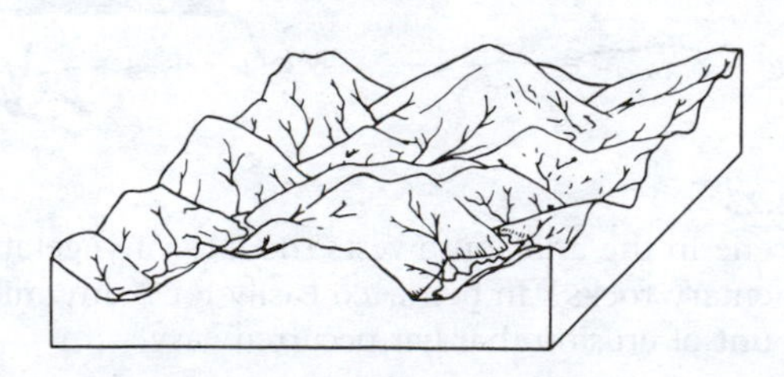

FIGURE 2.25
Mature stage.

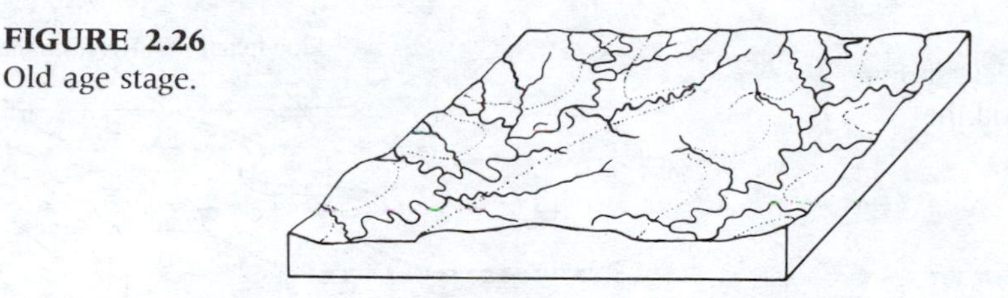

FIGURE 2.26
Old age stage.

progresses. Thus, there is no great change in shape, as there is in the Davis cycle of erosion, because once an equilibrium shape develops, no new shape can form unless the balance is upset by uplift or some other change.

In an effort to resolve this latter objection, a number of recent studies have been made of hillslopes in areas where the climate, vegetation, relief, and rock type are uniform. The slope angle of such hillsides is related both to stream gradient and to density of streams. Where direct measurement of slope erosion was made, it was found that during retreat of slopes, the slope angle remained constant if the debris was removed from the foot of the slope. If the erosional debris was not removed from the foot of the hill, the slope angle was reduced. The latter case is probably more common in humid areas where weathering may be deep and creep may be active. All of this suggests that the Davis cycle is not only a special case, but an oversimplification as well.

The length of time and the size of the area considered in the study of landscape also may influence one's ideas. A single hillside or stretch of river may be in balance during the time of a short study of the type described above; however, if the whole drainage area is considered over a geologically long time interval, erosion must remove appreciable amounts of mass and so change the system. Thus, in the short term, landforms may not seem to change with time, but over a long interval, they might actually do so. The dynamic earth may not remain still long enough for these changes to occur, however, for the earth's surface is the result of the constant struggle between erosion and uplift.

Floods

> **Floods are as much a part of the phenomena of the landscape as are hills and valleys: they are natural features to be lived with, features which require certain adjustments on our part. (W. G. Hoyt and W. B. Langbein, *Floods,* Princeton Univ. Press, 1955)**

Some primitive peoples adapted reasonably well to rivers and accepted seasonal or occasional floods as part of life. As people developed and built bigger and better structures for their comfort, they became less willing to accept the natural behavior of rivers.

A typical river valley consists of a channel, a flat area called a **floodplain**, and sloping valley sides (Fig. 2.27). Most of the time the flow is confined to the channel; but at times of high flow, when the channel cannot carry all of the water, it spills

FIGURE 2.27
A typical river valley, showing the channel and the floodplain.

onto the aptly named floodplain. The floodplain becomes the channel of the river during times of high water. This is the normal, expected behavior of a river. People who live or build structures on floodplains must expect these floods. Flood control is never 100 percent effective and actually can increase the damage, as we will see later. Floods are not entirely bad events because the silt deposited on the floodplains renews the soil of these generally very productive agricultural areas.

Any process that puts more water into a river than it can carry will cause a flood. Intense precipitation and rapid melting of snow, especially in mountainous headwaters, are common causes. Other natural processes, such as dams caused by landslides, can cause upstream floods. In this case, commonly when the lake behind the slide overtops the slide-dam, the rapidly moving water quickly erodes the slide, emptying the lake rapidly and causing downstream flooding. When the ice breaks up each spring on northern rivers, the floating ice-blocks themselves can cause damage to bridges and the like, and ice jams cause floods, first upstream and then downstream, in the same way as the slide dams just described. Except for landslides, the natural processes described are seasonal and are controlled by climatic processes.

Flood prediction requires much data and cooperation between hydrologists and meteorologists, and even at best it is not an exact science. Hydrologists must know such characteristics of the drainage basin as how long it takes for a rainfall of a given intensity to cause an increase in discharge of the river. This increase will occur after a low-intensity rain has saturated the ground, and almost at once in a high-intensity storm, but also will depend on how much moisture was in the soil at the start of the storm. Hydrologists also must know the characteristics of the river. How much discharge can it carry before it floods? How much area will be flooded as the height of water rises above bank-full? How rapidly will the wave of high water move downstream? Meteorologists must predict the length and intensity of the storm. If melting snow is involved, they must consider the many factors involved, such as temperature, humidity, and wind, as well as the amount and condition of the snow. A number of stream gauging stations that report stream height and discharge and weather stations that report precipitation and other weather elements are necessary to gather the needed data in a timely fashion.

Data on historical floods have been published by the U.S. Geological Survey in a series of publications called *Hydrologic Investigations Atlases*. These atlases show areas flooded, stream gauge data, and other data. Flood frequency data show the expected recurrence of floods of different height—that is, the size of flood that can be expected every 25 to 50 years, and so forth. If one of these atlases is available for your area, it should be consulted.

Urbanization, Floods, and Erosion

The main effect of urbanization is to increase runoff, which in turn causes increased erosion and, of course, deposition of the eroded material. The resulting flooding, erosion, and deposition, as we have seen, may also occur in urbanized areas, compounding the problems. In a given drainage basin, a rainfall will, after some time, cause an increase in the flow of the main stream. The length of the time lag and the amount of increase in the flow depend on the amount and intensity of the rainfall, as well as on the characteristics of the river. Figure 2.28 shows an example. Urbanization

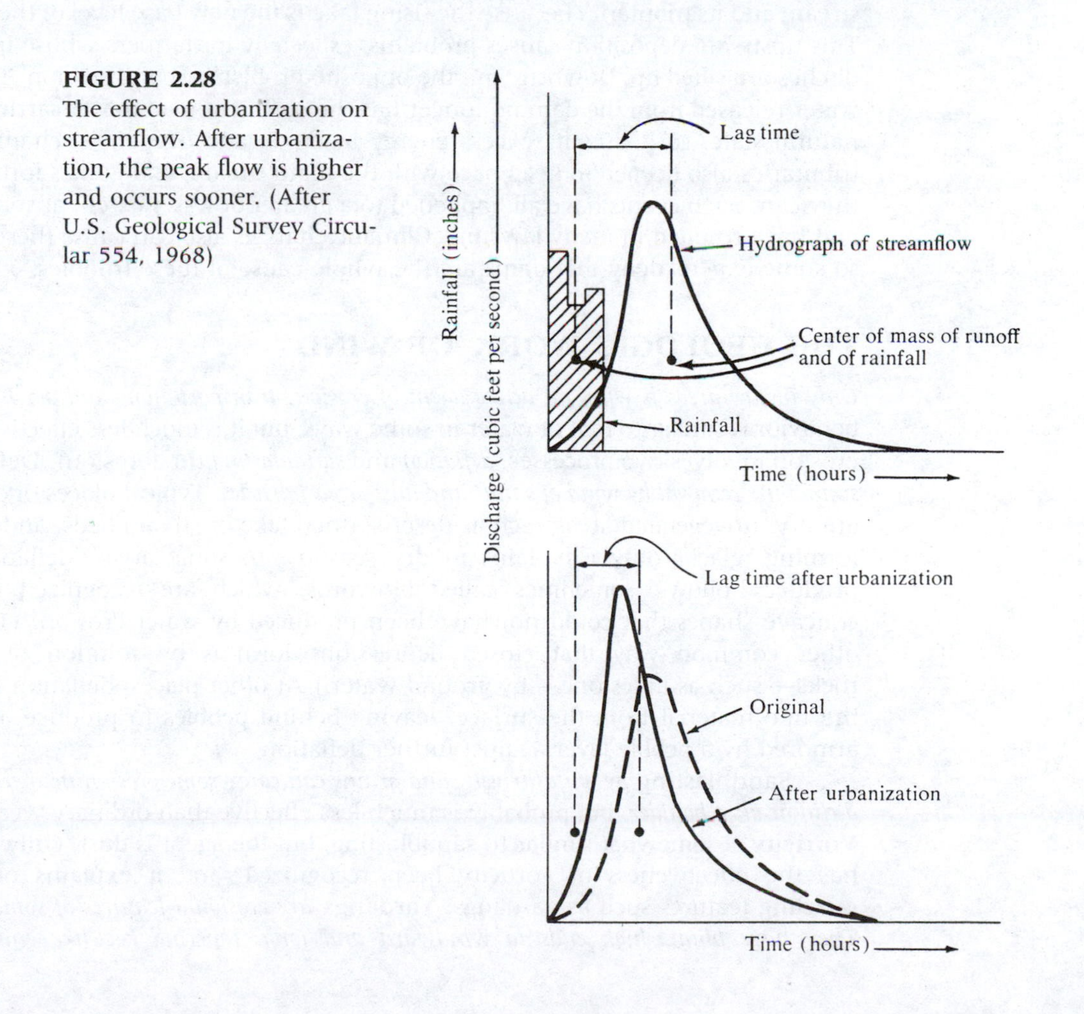

FIGURE 2.28
The effect of urbanization on streamflow. After urbanization, the peak flow is higher and occurs sooner. (After U.S. Geological Survey Circular 554, 1968)

of a drainage basin increases both the peak runoff and the total runoff because buildings and pavement reduce the infiltration.

The increase in discharge causes increased erosion. The first effect is a deepening of the channel and removal of accumulated debris, thus increasing the river's capacity so that it can carry more water without flooding. As the main channel deepens, the tributary streams also deepen because their base level, which is the elevation at which they join the main stream, is lowered. Gullies may appear at many places in the drainage basin.

The increased runoff means less infiltration and, therefore, less recharge of ground-water supplies. The water table will fall, and water wells must be drilled deeper.

A dam causes changes in a stream. Dams have finite lives because the sediment carried by the stream eventually fills the lake created by the dam. This deposition upstream from the dam occurs not only in the lake but all along the stream and its tributaries because the rising lake is the new base level of the stream. This upstream deposition causes problems, especially for farmers whose irrigation ditches are silted up. Downstream, the opposite problem occurs: erosion. The clear water released from the dam no longer has the sediment load that it carried in the natural state, so it uses its excess energy to erode and deepen its channel. The tributaries also deepen to keep pace with the main stream, and gullies form where they can. Such events have all happened to a greater or lesser extent at some dams and have resulted in many lawsuits. Climatic changes also can cause these effects, so some experts deny that dams are the whole cause of these troubles.

THE GEOLOGIC WORK OF WIND

Only in dry areas is wind an active agent of erosion, transportation, and deposition. Its behavior is similar to that of water in some ways, but it is much less effective. Wind erosion involves two processes, *deflation* and *sandblasting* (or abrasion). **Deflation** *is simply the removal by wind of sand- and dust-sized particles.* Typical places of deflation are dry, unvegetated areas such as deserts, dried lake or stream beds, and actively forming, glacial-outwash plains in dry seasons. In some areas, deflation may produce hollows, sometimes called blowouts, which are recognized by their concave shapes that could not have been produced by water erosion. (The only other common way that closed depressions form is by solution of soluble rocks—such as limestone—by ground water.) At other places, deflation removes the fine material from the surface, leaving behind pebbles to produce a surface armored by a pebble layer against further deflation.

Sandblasting *by wind-driven sand grains can cause some erosion near the base of a cliff or on a boulder,* but probably is much less effective than ordinary weathering. **Vorticity** is somewhat similar to sandblasting, but the agent is dust. Only recently has the effectiveness of vorticity been recognized, and it explains otherwise puzzling features such as yardangs. **Yardangs** *are erosional features of some deserts. They have blunt, high ends to windward and long, tapering keel-like ends in the*

FIGURE 2.29
Yardang near Kharga, Egypt. Wind comes from the right. The blunt windward end is eroded by sandblasting, and the tapering downwind part is shaped by vorticity. (Photo courtesy Marion I. Whitney)

downwind direction (Fig. 2.29). The high, blunt, wide windward ends are carved by sandblasting and the fluted keel is the product of vorticity. *Dust carried by spinning or eddying winds is the eroding agent of vorticity.* Surprising as it may seem, this process is capable of eroding hard rock. On a much smaller scale, vorticity produces similar shapes on pebbles called **ventifacts** or *dreikanter* (German for "three edges") (Fig. 2.30).

Wind transportation is similar to water transportation in that both a suspended load and a bed load are involved. The suspended load produces dust storms, and the bed load forms sandstorms. Most windblown sand is less than 1 mm in diameter, and the average size is about 1/4 mm. All material finer than 1/16 mm (silt) is considered dust. The separation into these two size ranges is very sharp, much more so than in the case of water transportation, so that wind-deposited sediments are generally much better sorted by size than are water-laid deposits. The reason for this better sorting is that wind can raise sand particles only a meter or so above the ground, although it can raise dust thousands of meters. At one place where measurements were made, 50 percent by weight of the sand was raised less than 13 cm (5 inches) above the ground, and 90 percent was raised less than 64 cm (25 inches), although some large fragments rose more than 3 m (10 feet). At this

FIGURE 2.30
Ventifacts. (Photo by M. R. Campbell, U.S. Geological Survey)

place, maximum abrasion occurred 23 cm (9 inches) above the ground.* Such abrasive effects (only a few centimeters above ground) are common on fence posts and power poles in areas of sandstorms.

As might be imagined, sandstorms are associated with dust storms, but dust storms may occur without sandstorms; so, in general, dust is moved much farther than sand. The mechanism for moving sand is similar, in part, to the action of water. The process produces a bouncing motion of the sand grains, which begins when a grain rolls into or over another grain. When the first grain hits, the impact transfers its kinetic energy to one or more grains which, in turn, also bounce up and upon falling, continue the process by hitting other grains (Fig. 2.31).

Wind deposits are of two types—**loess** *composed of dust,* and drifts and dunes composed of sand. *Loess forms thick, sheetlike deposits that are relatively unstratified because of their rather uniform grain size, generally in the silt size range.* Thick loess deposits in China are believed to have originated from the deserts of Asia. Most American loess deposits are associated with glacial deposits and are believed to

*R. P. Sharp, "Wind-driven Sand in Coachella Valley, California," *Geological Society of America Bulletin* 75 (1964): 785–804.

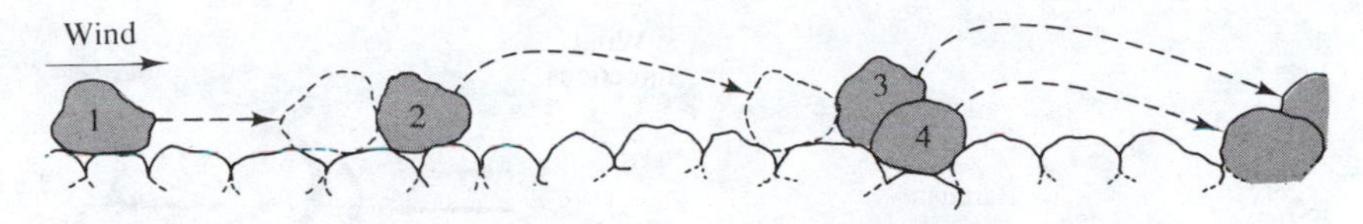

FIGURE 2.31
Movement of bed load. Sand grain 1 rolls over and hits grain 2, which bounces up and hits grains 3 and 4, both of which also bounce up and hit more sand grains.

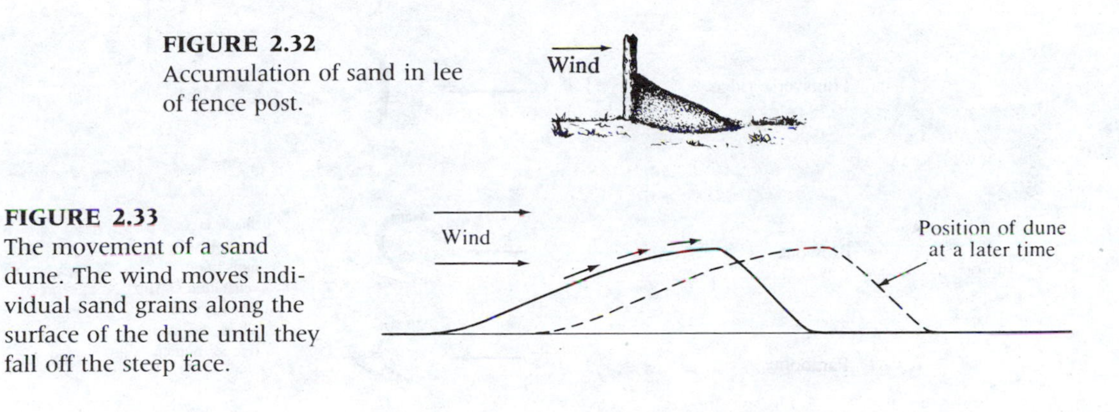

FIGURE 2.32
Accumulation of sand in lee of fence post.

FIGURE 2.33
The movement of a sand dune. The wind moves individual sand grains along the surface of the dune until they fall off the steep face.

have formed when glacial retreat left large unvegetated areas on which winds could act. Windblown sand, unlike loess, does not generally form blanket deposits. **Dunes** *are naturally formed accumulations of windblown sand.* Most dunes probably start in the lee of an obstacle (Fig. 2.32). Unless they have become fixed by vegetation, most sand dunes migrate slowly across the desert (Fig. 2.33). As might be expected, many dunes and dune areas are not easily classified; however, several distinct types of dunes are common in dry areas (Fig. 2.34).

THE GEOLOGIC WORK OF ICE—GLACIERS

Although glaciers are very much less important than rivers in worldwide erosion, they have shaped many of the landforms of northern North America. In addition, most of the mountain ranges of western America have been greatly modified by mountain glaciers.

A **glacier** *is a mass of moving ice.* Glaciers form as a result of accumulation of snow in areas where more snow falls than melts in most years. The accumulation of snow must, however, become thick enough so that it recrystallizes to ice. The recrystallization process depends on the pressure of the overlying snow, which transforms the light, loosely packed snow into small ice crystals; with higher pressure, the small crystals become larger. The process is quite similar to the metamorphism of sandstone into quartzite and limestone into marble. The tendency of snow to recrystallize and form larger crystals is well known to anyone

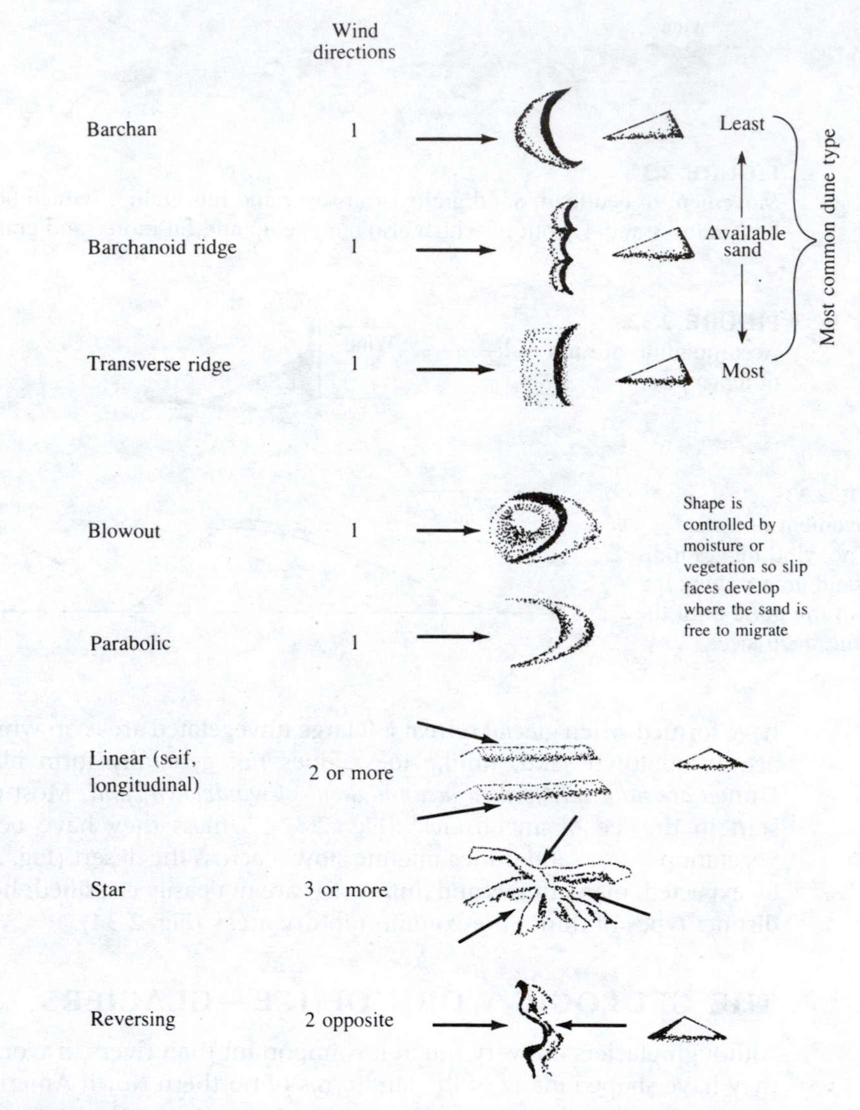

FIGURE 2.34
Classification of sand dunes.

who skis or who has lived in snowy areas. Falling snowflakes are generally small and light, but after a few sunny days or in the spring, they are much larger at the surface, and digging will reveal even larger crystals. When the amount of ice formed in this way becomes large enough that it flows under its own weight, a glacier is born. We normally see ice as a brittle substance, so the fact that a glacier flows under its own weight, much like tar, may seem strange; however, a mass of ice about one hundred meters thick behaves as a very viscous liquid.

Glaciers are of two main types: the *alpine* or *mountain* glacier and the much larger *continental* or *icecap* glacier.

Mountain Glaciers

Mountain glaciers develop in previously formed stream valleys because valleys are lower than the surrounding country and thus become accumulation sites for snow (Fig. 2.35). When enough ice has formed, the glacier begins to move down the valley. How far down the valley it extends will depend on how much new ice is formed and how much melting occurs each year. As the glacier extends farther down the valley, the amount of annual melting will increase until a point is reached where the amount of melting is equal to the amount of new ice added. Such a glacier is in equilibrium. If the amount of snowfall increases or the summers become cooler, it will advance farther; if the snowfall decreases or the summers become warmer, it will melt back. However, it generally takes a number of years for the more- or less-than-normal snowfall of a single year or group of years to become ice and to reach the snout of the glacier. This means that, although a glacier is

FIGURE 2.35
Mountain glaciers. Several mountain glaciers join in this valley. Note the crevasses on the surface of the ice and the moraines, both on the valley sides and separating the ice tongues from the different glaciers. (Photo from Swissair Photo Ltd.)

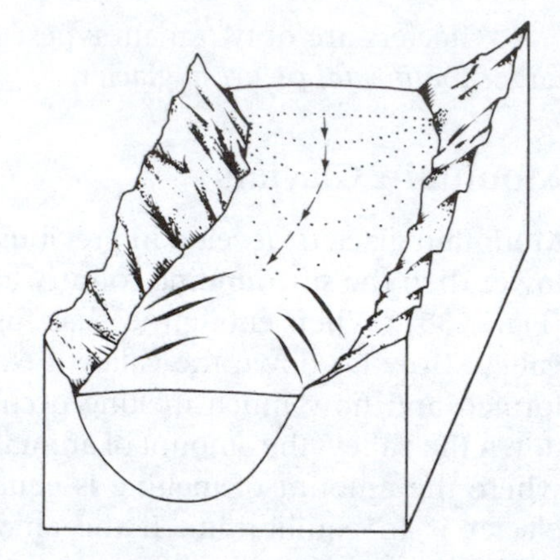

FIGURE 2.36
Movement of glaciers and the development of crevasses. The movement is shown by the successive locations of stakes driven into the ice in the upper part of the figure. The opening of crevasses in the brittle ice as a result of the greater movement of the center of the glacier is shown in the lower part of the figure.

sensitive to climatic changes, it tends to average snowfall and temperature changes over a number of years. Hence, the advance or the retreat of glaciers gives some information on long-term climatic changes. For example, study of old photographs and maps shows that the glaciers in Glacier National Park, Montana, (and elsewhere) have retreated since the turn of the century. About 1945, glaciers began to advance again. It is too soon to say whether this is a long-term change and whether this readvance will continue. One point needs emphasis: *The glacial ice is always moving down-valley, even if the snout of the glacier is retreating because of melting.*

The movement of glacial ice can be measured by driving a series of accurately located stakes into the glacier and surveying them periodically. As would be expected, because of friction, the sides of the glacier move more slowly than the center. The upper 30–60 m (100–200 feet) of the glacial ice behaves brittlely so that the different rates of flow at the surface open large cracks, called *crevasses,* in this brittle zone (Fig. 2.36). The movement of mountain glaciers varies from less than 2.5 cm (1 inch) per day to more than 15 m (50 feet) per day.

Erosion by mountain glaciers produces spectacular mountain scenery, such as Yosemite Valley in the Sierra Nevada of California. Glaciers erode by plucking large blocks of bedrock and by abrasion. **Plucking** *is accomplished by meltwater that flows into joints in the bedrock and later freezes to the main mass of ice which, on advancing, pulls out or plucks the block of loosened bedrock.* These blocks, together with debris that falls on the glacier from the valley sides, arm the moving ice with rasplike teeth that grind the sides and bottom of the glacial valley (Fig. 2.37). Glacial erosion greatly modifies the shape of the stream valley occupied by the glacier. Most of the erosion is probably done by plucking, and abrasion generally smooths and even polishes the resulting form.

A glaciated valley differs from a stream valley in that it is deepened, especially near its head, and the sides are steepened so that its cross section is changed from V*-shaped to* U*-shaped* (Fig. 2.38). Glacial erosion is most active near the head of the glacier, and the deepening there flattens the gradient of the valley. This deepening, together

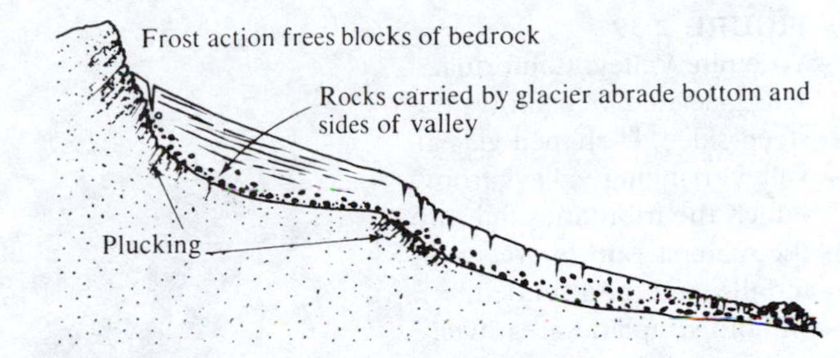

FIGURE 2.37
Longitudinal section through a mountain glacier, showing development of a cirque by plucking, and reshaping of the valley by plucking and abrasion. This sketch shows an early stage in the development of a glacial valley.

with the steepening of the sides and the head of the valley, produces a *large amphitheaterlike form that closely resembles a teacup cut vertically in half.* **Cirque** *is the name applied to this form.*

A glacially eroded valley is recognized by the U-shape, the cirque at the head, and the grooves, scratches, and polishing due to abrasion (Fig. 2.40C). In addition, any small hills or knobs in the valley have been overridden by the glacier, and they are rounded and smoothed by abrasion on the struck side and steepened by plucking on the lee side. Many glaciated valleys are nearly flat, with rises or steps where more resistant rocks crop out. A glacier moving down a valley also tends to straighten the valley, because the glacier cannot turn as abruptly as the original river could. This has the effect of removing the spurs or ridges on the insides of

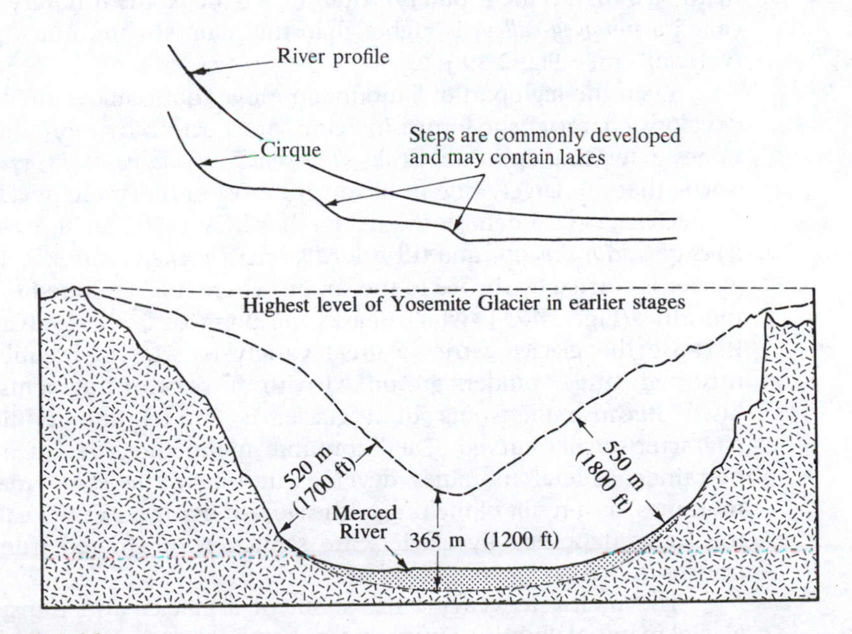

FIGURE 2.38
Glacial modification of river valleys.

FIGURE 2.39
Yosemite Valley, California. This glaciated area shows the steep-sided, U-shaped glacial valley, hanging valleys from which the tributaries fall into the main stream in great waterfalls, and the rounded, smoothed upland area over much of which the glaciers also moved. The present river is very small for the size of the valley. (Photo by Sarah Ann Davis, Courtesy California Division of Mines and Geology)

curves of the stream valley. The amount of ice in a main-stream valley, whose source is near the crest of the mountain range, is generally much greater than the amount of ice in a tributary-stream valley; hence, the main valley is eroded deeper by the ice than is the tributary. After the ice melts, the tributary stream (in a valley called a *hanging valley*) is higher than the main stream, into which it flows via a waterfall. (See Fig. 2.39.)

Even the high part of a mountain range that is above the level of the glaciers develops characteristic forms. Increased frost action here produces narrow, jagged ridges extending up to the peaks. *The peaks develop pointed, pyramidal shapes called* **horns** that are largely due to headward erosion by cirque development.

Glaciers also deposit the debris that they carry. **Drift** *is the name used for all types of glacial deposits,* and **till** *is used for ice-deposited sediments.* The material in the glacier is carried by the ice to the snout, where it is deposited to form the terminal **moraine** (Fig. 2.40A), which marks the point of farthest advance of the glacier. Because the glacier carries a great variety of sizes of fragments, moraines are unsorted; huge boulders are mixed with all sizes of fragments down to silt and finer. Because the snout of a glacier is curved, the terminal moraines are characteristically curved. They continue along the sides of a glacier as lateral moraines. Lateral moraines develop, in much the same manner as terminal moraines, as a result of melting of the slow-moving ice at the valley sides. Many of the rocks deposited by glaciers are scratched or striated from abrasion during transport.

The meltwater leaving the snout of a glacier also transports and deposits glacial material, forming outwash deposits that in some cases form thick sheets. The meltwater from an active glacier is colored white due to fine material, called **rock**

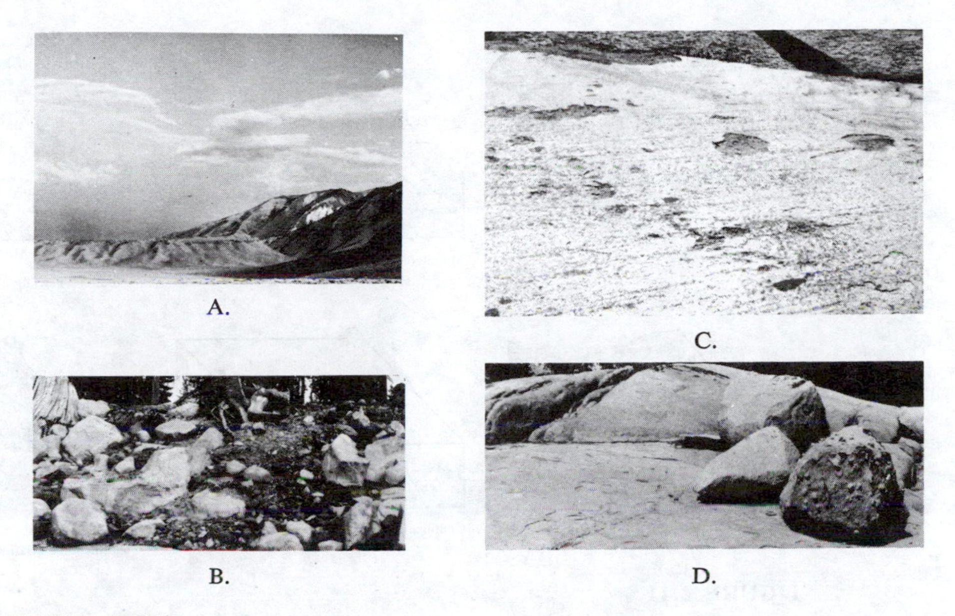

FIGURE 2.40
Glacial features. A. Terminal moraine. B. Material in a glacial moraine containing many kinds of rock in all sizes from boulders to silt. C. Polished and grooved granite caused by the rock-studded base of a moving glacier. D. Erratic boulders abandoned by a glacier when it melted away. (Photos by Mary Hill, Courtesy California Division of Mines and Geology)

flour, from the abrasive erosion of the glacier. This material resembles clay, but microscopic examination shows that it is composed of fine mineral fragments, not clay. (It is the source of the **varved "clays,"** to be discussed later.) When a glacier melts, its transported load is dropped wherever it happens to be and forms ground moraine. Blocks of ice buried in ground moraine, upon melting, leave depressions that may become small ponds.

Continental Glaciers

The geologic work of continental glaciers is very similar to that of mountain glaciers. Their erosive effects, however, are much less and are generally limited to rounding the topography and scraping off much of the soil and overburden (Fig. 2.41). Their deposits are like those of mountain glaciers, but are more extensive. One type of deposit that helped lead to the recognition of continental glaciation is the **erratic boulder** (Fig. 2.40D). This term is applied to *boulders, some of which are quite large, that are foreign to their present locations.* In some areas, occurrences of distinctive erratic boulders can be traced back to their origin, thus indicating the path of the glacier.

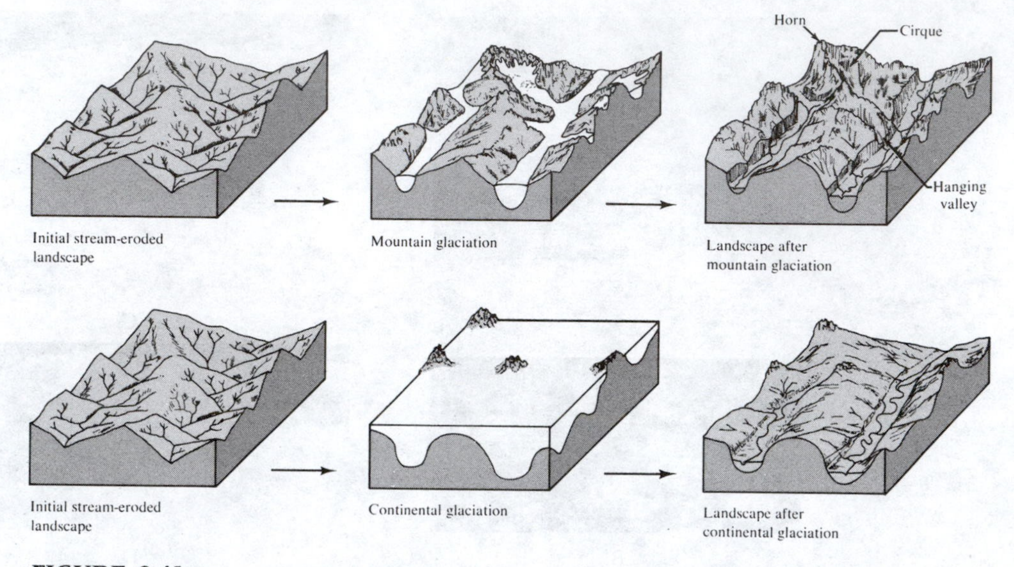

FIGURE 2.41
Mountain and continental glacial erosion of a hypothetical landscape are shown. Continental glaciers create smooth, rounded topography. Frost action above mountain glaciers forms jagged peaks.

Continental glaciers occupied parts of North America and Europe in relatively recent geologic time, and they occupy Greenland and Antarctica today. During the recent glacial period, there were four main stages of glacier development in both Europe and America. In North America there were several centers of ice accumulation in Canada. Glaciers moved out from these centers in all directions. Thus, although the ice moved generally southward in the United States, it moved northward in northern Canada. *In no sense did the last glacial age result because ice from near the north pole moved south.* Recognition of four advances depends largely on field observations. In some places fresh, unweathered moraines overlie weathered moraines, which, in turn, overlie more deeply weathered moraines. Tracing such relationships led to recognition of four main stages with the extents shown on the map in Figure 2.42.

The most recent continental glaciation of *Pleistocene* age probably began about three million years ago. This is estimated from the degree of weathering of the deposits, from the amount of soil produced between glacial advances, from study of fossils, and from other methods. The last glacial advance began about 30,000 years ago and retreated about 10,000 years ago. This is known fairly accurately from carbon-isotope studies. The carbon-isotope method uses the fact that organic carbon has a fixed ratio of carbon isotopes when formed, but this ratio changes after death of the organism due to radioactive decay of one of the carbon isotopes (carbon-14). The method is not suitable for material more than about 40,000 years old. (Radioactive dating methods will be discussed more thoroughly in Chapter 4.)

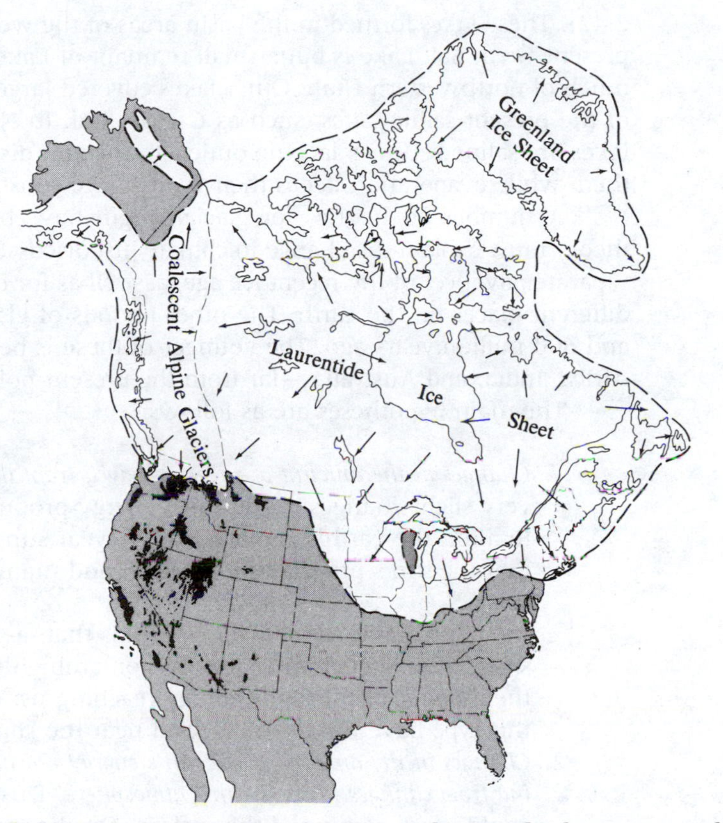

FIGURE 2.42
Maximum extent of Pleistocene ice sheets, and locations of the lakes of glacial times in western United States.

Before radiocarbon dating came into use, other methods were employed, particularly varved clays. This method is much like studying tree rings, as the varves are yearly deposits in glacial lakes. The dark part of each varve is deposited during the winter, and the light part in the summer. Thus, the thickness of varves records the climatic conditions, and the sequence of relative thickness is correlated from lake to lake until the life span of the glacier is covered. Then the total number of varves is counted. Apparently due to error in correlation, this method gave too high an estimate.

Cause of Glaciation

To discover the cause of glacial periods, we must first see under what conditions glaciers form. Average temperatures must be lower than at present (but not greatly lower), and precipitation must be high. These two conditions will cause the accumulation of snow that can form glaciers. The temperature must be low enough to ensure that precipitation will be in the form of snow, but not so low as to inhibit precipitation. This latter point is illustrated by the lack of glacier formation, because of low precipitation, in the many very cold regions today. The heavy precipitation during at least parts of the glacial age is shown by the huge lakes that formed in areas south of the glaciers, where it was too warm for snow to accumulate (Fig.

2.42). These lakes formed in the basin areas of the western United States, and the present Great Salt Lake is but a small remnant of Lake Bonneville, which covered much of northwestern Utah. Other lakes covered large areas in Nevada, and some of the present saline lakes, such as Carson Sink in Nevada, are remnants. These lakes are saline because, lacking outlets to the sea, dissolved material builds up in them while evaporation keeps their levels fairly constant.

A number of theories for glacial origin have been proposed. A successful theory must explain the change in climate just discussed, and must account for the separate advances of the recent ice age, as well as for other periods of glaciation at different places on the earth. The other periods of glaciation occurred about 275 and 600 million years ago. The younger of these is better known and occurred in Africa, India, and Australia—far from the present poles.

The main hypotheses are as follows:

1. *Changes in the amount of energy received from the sun.* Astronomers detect very slight changes in the sun's energy production, very much too small to affect the earth's climate. The 11-year sunspot cycle, however, shows that the sun's processes do change, and many other stars are variable stars.

   A variation of this theory holds that, as the sun moves through space, it may encounter regions containing dust. Such dust would limit the amount of the sun's energy reaching us. However, dust clouds of this type have not been detected near the sun.
2. *Changes in the amount of the sun's energy reaching the earth's surface, resulting from changes in the earth's atmosphere.* Dust from volcanic eruptions could cause cooling of the surface. On the other hand, dark particles in the atmosphere might absorb the sun's energy and so heat the atmosphere. Volcanic dust does eventually settle to the earth's surface, where it forms thin layers of volcanic ash. The number of these layers can be counted in sedimentary cores recovered from deep-ocean sediments, and the ages of the layers can be determined by radioactive methods of dating. The results of such studies are inconclusive, and few believe that volcanic dust is the main cause of ice ages.

   Decreasing the amount of carbon dioxide in the atmosphere would cool the earth, as this gas allows radiant energy from the sun to pass through but prevents radiation from the earth from leaving. Thus, the carbon dioxide in the atmosphere acts much like a greenhouse. However, it is doubtful that the necessary changes in the amount of carbon dioxide are possible.
3. *Continental drift caused by plate tectonics can move continents nearer to the poles and so help to start an ice age.* At high latitudes temperatures are lowered, and this would aid, but not cause, development of glaciers. Continents near the poles would tend to obstruct poleward movement of warm tropical ocean water. Mountain building and uplift in parts of the continents also would cause locally cooler temperatures and would

affect atmospheric circulation. Continents probably must be near the poles for a glacial age to occur, but some other cause also must be in operation.

4. *Periodic changes in the earth's motion around the sun affect the amount of the sun's energy received at any point on the earth.* This is an old theory recently revived. In the 1930s Milutin Milankovich, a Serbian mathematician, suggested that the effects of the earth's motions could cause glaciers to form. His calculations were based on ideas that were nearly 100 years old at that time.

   Milankovich recognized three periodicities in the earth's movements: (a) The earth's orbit around the sun is not quite circular, and the average distance between sun and earth changes slightly, with about a 105,000-year period. The calculated effect of this period is very small because the total energy input from the sun only varies about 0.1 percent. (b) The seasons are caused by the tilt of the earth's rotational axis relative to the plane of the earth's orbit. The tilt is currently 23.5 degrees, but it varies between about 22 degrees and 24.5 degrees, with a period of 41,000 years. The changing tilt angle changes the contrast between the seasons. (c) The third periodicity is in the direction in which the earth's axis points. This motion is known as the *precession of the equinoxes,* and the period of this motion is about 26,000 years. The precession determines where on the orbit the seasons occur, that is, whether winter in the northern hemisphere occurs when the earth is closest, or farthest, or some other distance from the sun. This motion also accentuates the contrast among seasons.

   The effects of these cycles can be tested in several ways. Cores of sedimentary rocks that range in age from the present back through much of the last ice age can be obtained in some lakes and in the deep ocean. The pollen in the cores obtained on the continents can be studied, and climatic changes can be inferred from the types of trees and plants. In marine cores, tiny fossils also can be studied, and the temperature of the sea water can be determined from study of the isotopes of oxygen present.

   The cores that have been studied revealed climatic cycles of 23,000 years, 41,000 years, and 100,000 years, in remarkably close agreement with Milankovitch's theoretical predictions. This suggests that these orbital motions do affect climate; however, in detail, the agreement is not so good. The 100,000-year cycle is dominant in the cores, and the theory predicts that it should have the smallest effect. There also is a lag of 7000 to 8000 years between the predicted and actual time of maximum extent of glaciation. Finally, the glacial advances in the last ice age lasted between 15,000 and 30,000 years, not the 10,000 predicted.

5. *Changes in the circulation of the oceans.* This is a promising hypothesis that seems adequate for the recent glacial periods. The hypothesis begins with an ice-free Arctic Ocean. With the present pole positions and our

present atmospheric circulation, the ice-free Arctic Ocean would cause the heavy precipitation necessary to initiate a glacial advance. By the middle of the glacial advance the Arctic Ocean would freeze, cutting off much of the supply of moisture. Precipitation would continue until the North Atlantic became too cold to provide enough moisture. Melting would then begin and continue until the Arctic Ocean was again ice-free. The stage is now set for the cycle to begin again.

This hypothesis can account for glacial periods only when the poles are located as they are now. If both poles were over open ocean, then the atmospheric circulation could not develop glacial periods. So, in order to initiate a glacial period, the poles would have to shift. Thus, this theory accounts for the four glacial advances of the recent glacial period; it also can explain the irregular periods of older glaciation. Recent studies, however, suggest that the present Arctic ice may be much older than the last glacial advance, a finding which casts doubt on this theory.

All of the causes of glaciation discussed have both advocates and dissenters because none of the hypotheses is entirely satisfactory.

It should be obvious from this discussion that the climatic conditions causing glaciation also had far-reaching effects, even far from the areas of actual glacial accumulation. It is estimated that sea level was lowered about 150 m (500 feet) during the glacial advances, and that if the present-day glaciers melt, sea level will rise between 30 and 45 m (100 and 150 feet).

GROUND WATER

It was not until the latter part of the seventeenth century that the origin of river water was shown to be rainfall. When this measurement was undertaken, it was discovered that only a fraction of the amount of water that fell as rain on a drainage basin flowed out as the river. The fate of the rest of the precipitation is not hard to imagine. Some of it evaporates, some is used by plants, and some seeps into the ground. The latter is the origin of the ground water that forms springs and wells.

The surface below which rocks are saturated with water is called the **water table.** This level changes with the season. In general, it is a reflection of the surface topography but is more subdued—that is, has less relief—than the surface topography (Fig. 2.43). Lakes and swamps are areas where the land surface is either below or at the water table. **Springs** occur where the water table is exposed, as on a valley side. The position of the water table changes seasonally, explaining why some springs are dry in summer. Springs may form where the water table is **perched;** *this situation results where the downward percolation of rainwater is stopped by a relatively impermeable rock such as a shale.* Springs of this type may flow only during the wet season (Fig. 2.43).

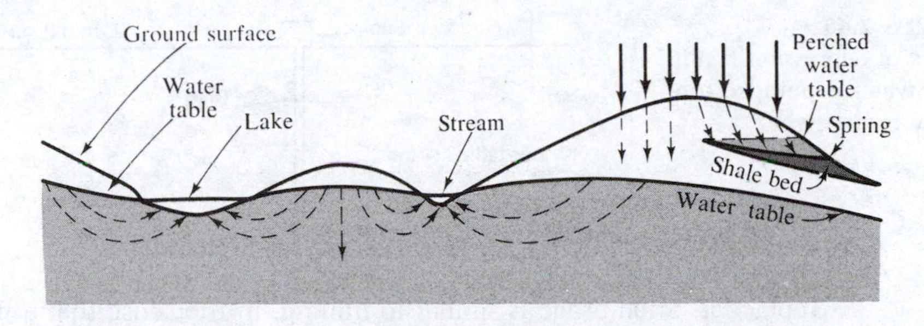

FIGURE 2.43
The water table has less relief than the topography. Arrows show the movement of ground water.

The action of ground water is controlled by the physical properties of the rocks. The amount of water that can be stored is determined by the amount of pore, or open, space in the rocks. The availability of the water is determined by the interconnections of the pore space, which is, of course, the *permeability* (Fig. 2.44). The most common reservoir rock is sandstone, although fractured granite or limestone, as well as many other rock types, can serve equally well.

Ground water is a very valuable commodity. It is recovered from wells for domestic, industrial, and agricultural use. In areas where ground water is used extensively, care must be taken that, on the average, no more water is withdrawn in a year than is replaced by natural processes. This can be determined by seeing that the water table is not lowered. The conservation of ground water is important because it moves very slowly, and many years may be required to replace hastily pumped water. The average rate of movement of ground water through rocks is only about 15 m (50 feet) per year, although at some places the movement is much faster.

This slow movement of ground water requires another caution. *If water is pumped too rapidly from a well, the water is removed from the immediate vicinity of the well faster than it can be replaced. The result is called a* **cone of depression** (Fig. 2.45). If cones of depression are developed around several closely spaced wells, the water table can be drastically lowered. Thus, the spacing and the rate of pumping of wells must be regulated to ensure the most efficient use of the water. At some places water has been withdrawn so rapidly that hundreds of years may be required to

FIGURE 2.44
The sorting of sandstone affects its porosity and permeability.

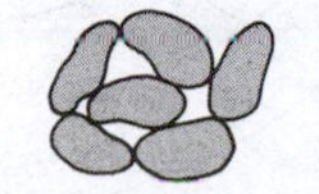
Well-sorted sandstone. Large amount of pore space gives high porosity and permeability.

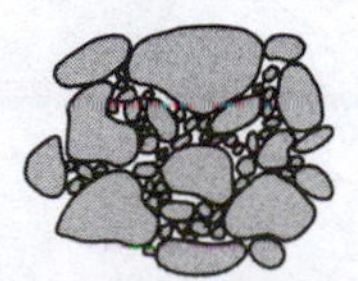
Poorly sorted sandstone. Has much lower porosity and permeability.

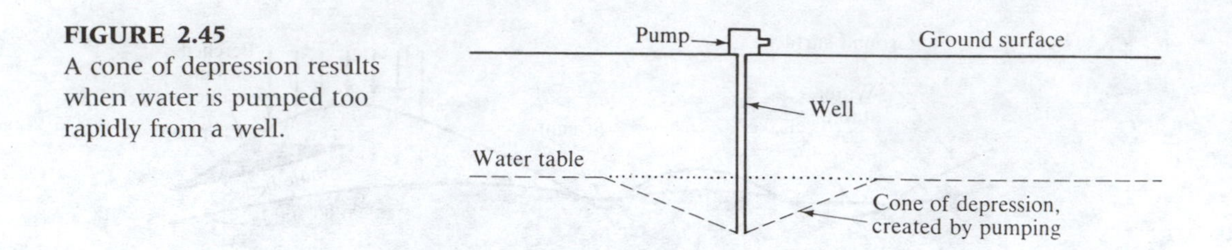

FIGURE 2.45
A cone of depression results when water is pumped too rapidly from a well.

replace it. Such usage is similar to mining, in the sense that a natural resource is exhausted.

A very important type of well is the **artesian well** (Fig. 2.46). In an artesian well, the reservoir is charged at a place where it is exposed, and the water is forced to move through the reservoir, which is confined above and below by impermeable rocks. Because water seeks its own level, a well drilled into an artesian reservoir may flow without pumping. However, the resistance to flowage of the water through the reservoir prevents the water from flowing to the same elevation as its recharging area; therefore, flowing wells must be lower in elevation than the recharge area (Fig. 2.46).

Both surface disposal and burial of wastes pose the danger of contaminating water, especially underground water. In times of flooding or heavy precipitation, surface waters also can be polluted. Decomposing garbage in a landfill produces methane, carbon dioxide, ammonia, and hydrogen sulfide. The carbon dioxide combines with water, forming carbonic acid, which increases the ability of the water to dissolve substances. Therefore the amount of soluble materials in the water is increased. Such water generally pollutes the surface or ground water into which it seeps. However, it is possible to avoid pollution from disposal sites by paying attention to the ground-water conditions, as shown in Figure 2.47.

Ground water forms caves, but exactly how is not clear (Fig. 2.48). Caves are formed mainly in limestone, but can form in any soluble rock. Calcite is soluble in water, especially in water that has carbon dioxide dissolved in it, as noted in the discussion of chemical weathering. It is probable that caves form in the zone of seasonal water table fluctuation by solution of limestone along joint planes. A

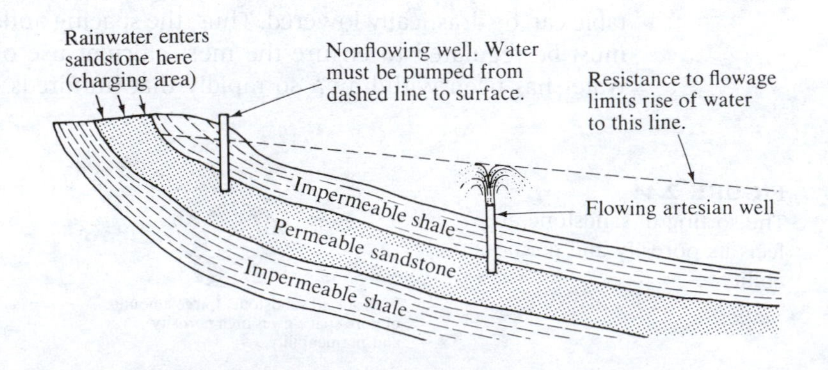

FIGURE 2.46
Artesian water results when the flow of ground water is restricted by impermeable layers.

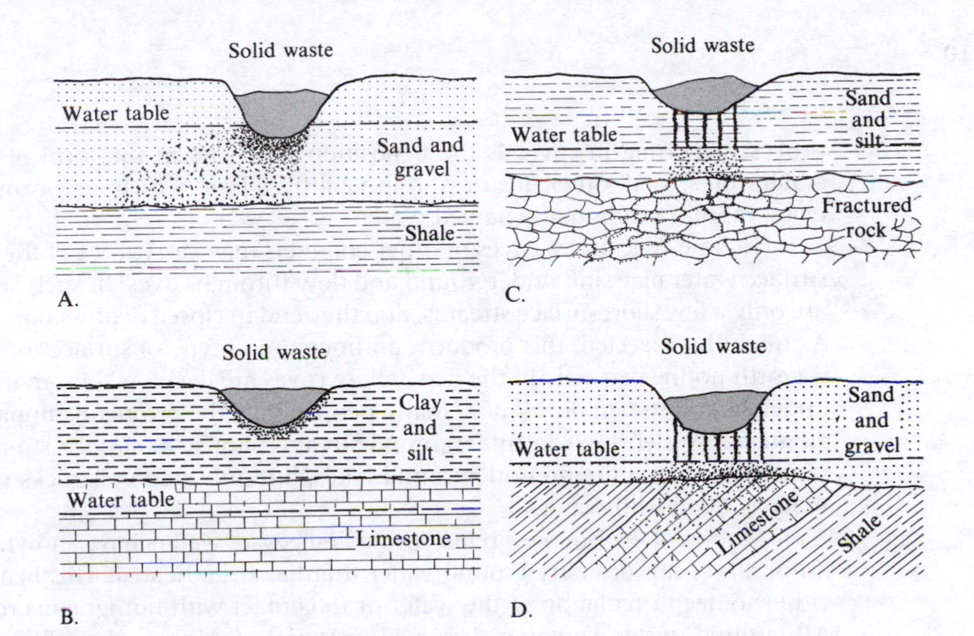

FIGURE 2.47
Effects of waste disposal on ground water. Only in B is contamination prevented by impermeable rocks. In all the other examples, contaminants from the waste move through permeable rock to the water table. (After U.S. Geological Survey Circular 601 F, 1970)

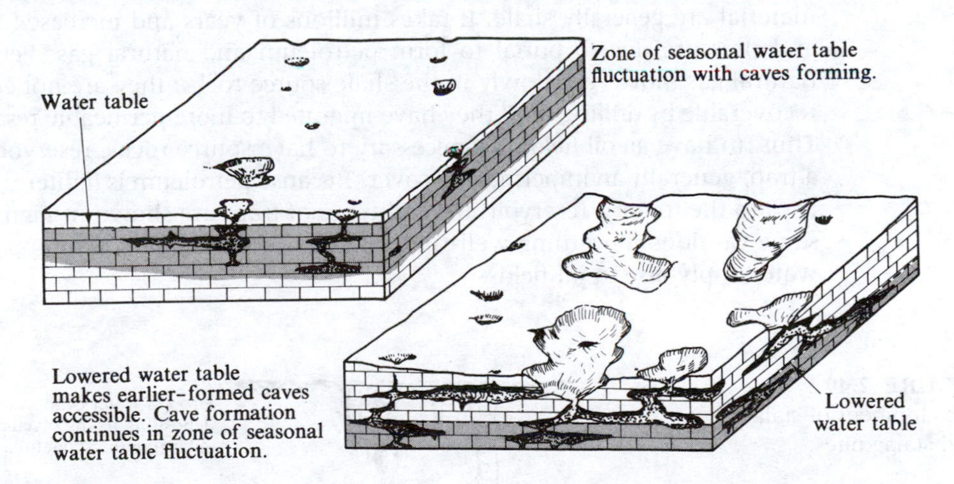

FIGURE 2.48
Development of caves. The caves begin by solution of limestone in the zone of seasonal water table fluctuation. Lowering of the water table makes the caves accessible. An area of cavernous limestone generally has few if any surface streams.

process occurring in caves is the deposition of calcite in the form of **stalactites, stalagmites,** and other features (Fig. 2.49), all of which are caused by the evaporation of carbonate-charged water.

In limestone areas, caves or other channels may carry most of the water. The surface water may sink underground and flow through caves. In such an area there are only a few short surface streams, and they end in closed depressions called *sinks*. As might be expected, this produces an uncommon type of surface topography. It is worth noting that only in the case where caves are developed is anything like an underground stream developed; many people have the erroneous impression that all ground water flows in underground streams similar to surface streams. In the same sense, there are no underground lakes, but only reservoir rocks whose pore space is filled with water.

Studies of the isotopes in hot spring and geyser waters have shown that most, if not all, of it is ordinary ground water that has been heated. This heating is due either to deep circulation of the water or to contact with hot igneous rock bodies. All ground water contains dissolved material. Spring and well waters have characteristic tastes due to this dissolved material, and the water from some springs and wells contains too much dissolved material to be useful.

Much of the foregoing information on ground water is applicable to the other natural fluids—petroleum and natural gas. These materials originate mainly from organic material in marine sedimentary rocks, although some do form in certain lake-deposited sedimentary rocks. The beds that contain the oil-forming organic material are generally shale. It takes millions of years and increased temperature and pressure due to burial to form petroleum and natural gas. Petroleum and natural gas move very slowly in the shale source rocks; they are not economically recoverable by drilling until they have migrated to more permeable reservoir rocks. Thus, to have an oil field, it is necessary to have source rocks, reservoir rocks, and a trap, generally an impermeable cover. Because petroleum is lighter than water, it rises to the top of a reservoir. Several types of traps are shown in Figure 2.50. The same cautions concerning well-spacing and rate of pumping that apply to ground water apply also to oil fields.

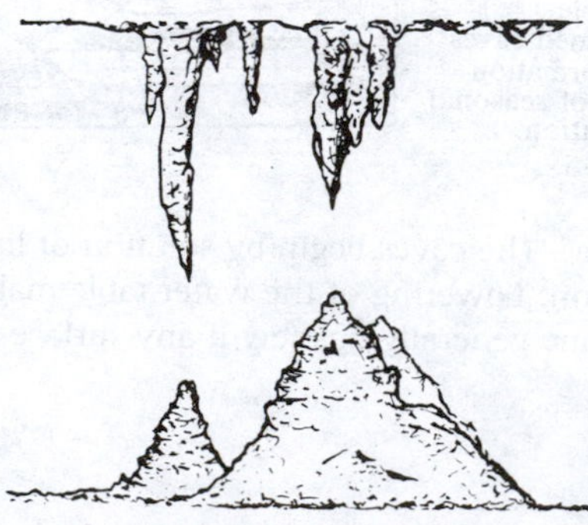

FIGURE 2.49
Development of stalactites and stalagmites.

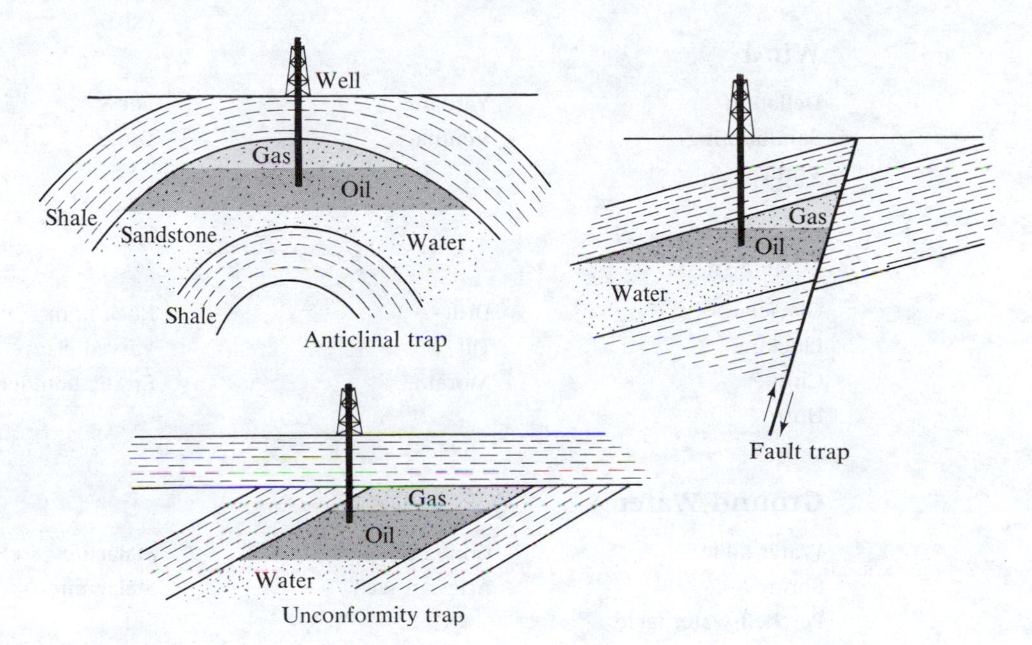

FIGURE 2.50
Several types of oil and gas traps. The terms anticline, unconformity, and fault are discussed in later chapters.

KEY TERMS

Downslope Movement

Uniformitarianism	Creep	Landslide
Overburden	Earthflow	Rockslide
Rockfall	Mudflow	Slump
Talus	Solifluction	

Rivers

Graded river	Mature valley	Terrace
Base level	Old age valley	Levee
Solution	Bed load	Youthful landscape
Abrasion	Stream piracy	Mature landscape
Pothole	Meander	Old age landscape
Dissolved load	Braided stream	Peneplain
Suspended load	Delta	Floodplain
Youthful valley	Alluvial fan	

Wind

Deflation
Sandblasting
Vorticity
Yardang
Ventifact
Loess
Dune

Glaciers

Glacier
Plucking
Cirque
Horn
Drift
Till
Moraine
Rock flour
Varved clay
Erratic boulder

Ground Water

Water table
Spring
Perched water table
Cone of depression
Artesian well
Stalactite
Stalagmite

QUESTIONS

1. Erosion tends to level the land surfaces. Why are the continents not worn flat?
2. Briefly describe creep.
3. Briefly describe rockslides.
4. How might rapid massive (downslope) movements be triggered? What is the role of water?
5. What are the processes of river erosion?
6. Describe the sequence of events in the formation and migration of a curve in a river course (use diagrams).
7. What are the three ways in which a river transports material?
8. A placer deposit is a place where a river has deposited fragments of heavy minerals such as gold. Where would you look in a river for such deposits?
9. What is a graded river?
10. The lower limit of stream erosion is called ________________ .
11. Show by a series of sketches the stages of erosion of a river valley.
12. What is the final surface produced by long-term erosion called?
13. Discuss whether building should be allowed on the floodplains of rivers.
14. How does urbanization affect the runoff of streams?
15. Briefly describe two kinds of wind erosion.
16. Wind-deposited silt is called ________________ .
17. Deposits of windblown sand are called ________________ .
18. Describe the sorting of wind deposits.
19. How might wind-deposited sediments differ from water-deposited sediments?
20. What is a glacier?

21. The two main types of glaciers are ________________ and ________________.
22. What changes does a mountain glacier produce in its valley?
23. How does a present-day glacier show climatic changes?
24. Why do crevasses form on the surface of glaciers?
25. Describe the processes of glacial erosion.
26. How do glacially deposited rocks differ from those found in river deposits?
27. What are the erosional effects of continental glaciers?
28. What is an erratic boulder?
29. How many glacial stages are recognized in the Pleistocene?
30. Which way did the ice move in northern Canada? What is the evidence?
31. What type of climate is most favorable for the development of glaciers?
32. What evidence do we have as to the climate during glacial times in southwestern United States?
33. List some suggested causes of continental glaciation.
34. Outline the theory of glaciation that depends on changes in ocean currents between the Arctic and the Atlantic oceans.
35. Which of the three agents of transportation—wind, water, and ice—is able to transport the largest boulders? Which is most limited as to the largest size it can transport?
36. How do mechanical weathering and glacial erosion work together?
37. What is the general relationship between the water table and the topography?
38. What is porosity? Permeability?
39. What happens near a well that is pumped too fast?
40. The average rate of movement of ground water is ________________.
41. Sketch an artesian well. How long ago did the water in such a well fall as rain?
42. Sketch several types of oil traps.
43. How are caves formed?

SUPPLEMENTARY READING

Downslope Movement

Dolan, Robert, and H. G. Goodell. "Sinking Cities." *American Scientist* 74(1) (January–February 1986): 38–47.

Voight, Barry, ed. "Natural Phenomena." Part 1 of *Rockslides and Avalanches.* New York: Elsevier, 1978, 834 pp.

Rivers

Carter, L. J. "Soil Erosion: The Problem Persists Despite the Billions Spent on It." *Science* 196(4288) (April 22, 1977): 409–11.

Curtis, W. F., J. K. Culbertson, and E. B. Chase. *Fluvial-Sediment Discharge to the Oceans from the Conterminous United States.* U.S. Geological Survey Circular 670. Washington: U.S. Government Printing Office, 1973, 17 pp.

Davis, W. M. "Physiographic Essays." Part II of *Geographical Essays.* New York: Dover Publications, (1909) 1954, 777 pp. (paperback).

Flemal, R. C. "The Attack on the Davisian System of Geomorphology: A Synopsis." *Journal of Geological Education* 19(1) (January 1971): 3–13.

Leopold, L. B., and W. B. Langbein. "River Meanders." *Scientific American* 214(6) (June 1966): 60–70.

Leopold, L. B., M. G. Wolman, and J. P. Miller. *Fluvial Processes in Geomorphology.* San Francisco: W. H. Freeman & Co., 1964, 504 pp.

Pillsbury, A. F. "The Salinity of Rivers." *Scientific American* 245(1) (July 1981): 54–65.

Wind

Greeley, Ronald, and J. D. Iverson. *Wind as a Geological Process.* Cambridge: Cambridge Univ. Press, 1985, 333 pp.

McKee, E. D. *A Study of Global Sand Seas.* U.S. Geological Survey Professional Paper 1052. Washington: U.S. Government Printing Office, 1979, 429 pp.

Péwé, T. L. ed. *Desert Dust: Origin, Characteristics, and Effect on Man.* GSA Special Paper 186. Boulder, CO: Geological Society of America, 1982, 303 pp.

Whitney, M. I. "Yardangs." *Journal of Geological Education* 33(2) (March 1985): 93–96.

Glaciers

Covey, Curt. "The Earth's Orbit and the Ice Ages." *Scientific American* 250(7) (February 1984): 58–66.

Denton, G. H., and S. C. Potter. "Neoglaciation." *Scientific American* 222(6) (June 1970): 101–10.

Huber, N. K. *The Geologic Story of Yosemite National Park.* U.S. Geological Survey Bulletin 1595. Washington: U.S. Government Printing Office, 1987, 64 pp.

Hutter, Kolumban. "Glacier Flow." *American Scientist* 70(1) (January–February 1982): 26–34.

Imbrie, John, and K. P. Imbrie. *Ice Ages: Solving the Mystery.* Short Hills, NJ: Enslow, 1979, 224 pp.

Koteff, Carl, and Fred Pessl, Jr. *Systematic Ice Retreat in New England.* U.S. Geological Survey Professional Paper 1179. Washington: U.S. Government Printing Office, 1981, 20 pp.

Radok, Uwe. "The Antarctic Ice." *Scientific American* 253(2) (August 1985): 98–105.

Ground Water

Dunne, Thomas, and L. B. Leopold. *Water in Environmental Planning.* San Francisco: W. H. Freeman & Co., 1978, 818 pp.

Heath, R. C. *Basic Ground-Water Hydrology.* U.S. Geological Survey Water-Supply Paper 2220. Washington: U.S. Government Printing Office, 1983, 84 pp.

Heath, R. C. *Ground-water Regions of the United States.* U.S. Geological Survey Water-Supply Paper 2242. Washington: U.S. Government Printing Office, 1984, 78 pp.

Issar, Arie. "Fossil Water Under the Sinai-Negev Peninsula." *Scientific American* 253(1) (July 1985): 104–13.

Keefer, W. R. *The Geologic Story of Yellowstone National Park.* U.S. Geological Survey Bulletin 1347. Washington: U.S. Government Printing Office, 1971, 92 pp.

Keller, W. D. "Drinking Water: A Geochemical Factor in Human Health." *Geological Society of America Bulletin* 89(3) (March 1978): 334–36.

Vogt, E. Z., and Ray Hyman. *Water Witching, U.S.A.* Chicago: Univ. of Chicago Press, 1959.

Caves

Ford, T. D., and C. H. D. Cullingford. *The Science of Speleology.* New York: Academic Press, 1976, 593 pp.

Moore, G. W., and G. N. Sullivan. *Speleology: The Study of Caves.* 2d ed. Teaneck, NJ: Zephyrus Press, 1978, 150 pp.

3 Structural Geology

INTERNAL STRUCTURE OF THE EARTH

Knowledge of the earth's interior cannot be obtained by direct observation and therefore comes from very diverse kinds of data. Astronomical observations reveal the size and shape of the earth and indicate that the interior is much different from the surface. Study of earthquake vibrations reveals much of the internal structure, and meteorites provide clues to the possible internal composition. The use of data from such diverse fields shows why the geologist's interests and training must be so broad.

Distribution of Mass in the Earth

The distribution of mass in the earth determines its motion in space. One can imagine two spheres of equal size and weight; one has a uniform composition, but the second has its weight concentrated near the center with the rest hollow, except for spokes that separate the shell from the core. These two spheres will have different mechanical behavior. If we roll them down an inclined plane, they will travel at different rates, the uniform sphere moving more slowly.

The earth's distribution of mass is determined by study of the earth-moon system. The moon's gravitational attraction on the earth's equatorial bulge causes the earth's axis to wobble very slowly. This motion is called the *precession of the equinoxes* and is quite similar to the wobbling motion of a spinning top as it slows down. (Precession is a term used to describe this type of wobble, and equinox is a term that designates a relationship between the earth's position and the "fixed" stars on certain days. Equinox ("equal night") occurs twice a year, at which times day and night are of equal duration.) This wobble requires 25,735 years to return to the same point, and it accounts for the fact that early Egyptian astronomers noted a different star in the position of Polaris, which is our present north star.

Because the amount of wobble (precession) caused by the moon's attraction depends on the mass distribution in the earth, it can be used to calculate the distribution of mass in the earth. This method also reveals that the earth is much denser at deep levels than at shallow levels.

The average specific gravity of the earth is about 5.5; that is, on the average, a given volume of it weighs 5.5 times as much as does the same volume of water. This is surprising because the rocks at the surface have average specific gravities of less than 3. Thus, the rocks at depth must have much greater specific gravities than those at the surface. This suggests that the composition of the deeper parts of the earth is very different from that of the crust.

Earthquakes and Seismology

Seismology *is the study of vibrations in the earth caused by earthquakes or explosions.* **Earthquakes** *are caused by sudden ruptures in the earth,* and the speed at which the resulting vibrations or waves travel through the earth provides some clues about the rock types. Figure 3.1 illustrates the probable origin of earthquakes, and Figure 3.2 shows how they are detected.

Three types of waves are generated by an earthquake (Fig. 3.3). They are:

1. **P waves**— primary or push waves. "Primary" because they travel fastest and so arrive first at a seismic station; "push" because they vibrate as compressions and rarefactions.

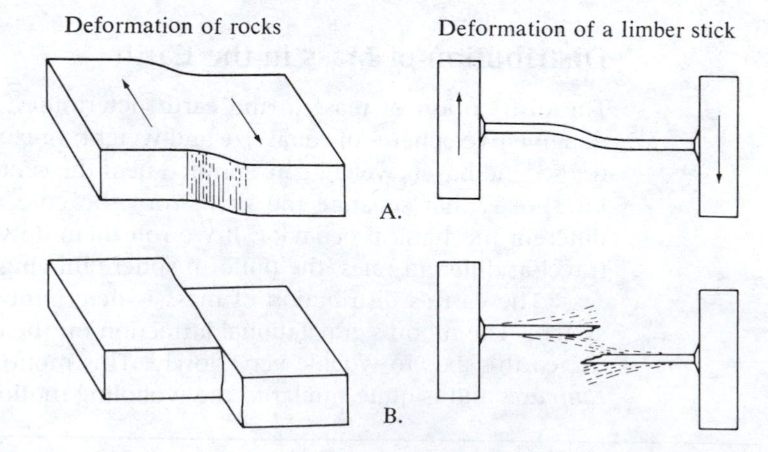

FIGURE 3.1
Origin of earthquakes. A portion of the earth's crust is shown on the left and a limber stick on the right. A. Slow deformation of the crust is caused by internal forces. B. When the strength of the rocks is exceeded, they rupture or fault, producing earthquake vibrations. Earthquakes on old faults result when the friction along the plane of the old break is exceeded.

2. **S waves**— secondary or shake (shear) waves. "Secondary" because they are the second arrivals at a seismic station; "shake" because they vibrate from side to side.
3. **Surface waves** — slow-moving waves with motion similar to the waves caused by a pebble tossed into a pond.

The propagation, or movement, of these waves illustrates more of the differences among them. The energy of an earthquake travels in all directions away from the source. The travel of the P waves is easily demonstrated: if a part of a coil spring is compressed and then quickly released, the resulting wave moves along the spring by successively compressing and spreading the spring. The S-wave movement is shown by displacing one end of the spring and then releasing. The

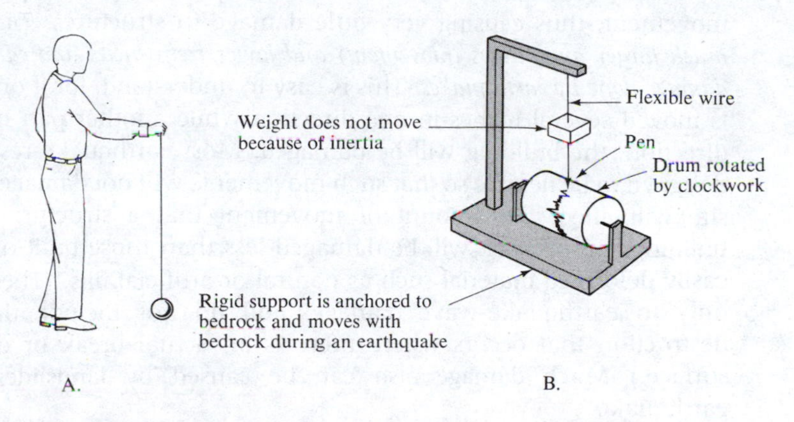

FIGURE 3.2
The operation of a seismograph. A. A weight on a string illustrates the principle of the seismograph. If the person's arm moves rapidly side-to-side, the weight remains still. B. In this diagrammatic model of a seismograph, the support is attached to bedrock and moves when the earth quakes. The weight tends to remain still, so the pen records the relative movement between the chart that moves with the bedrock and the weight that does not move. Actual seismographs are designed to respond to movement in a single direction. Instead of using a pendulum, the weights are suspended like doors and respond only to horizontal movements at a right angle to the plane of the weight.

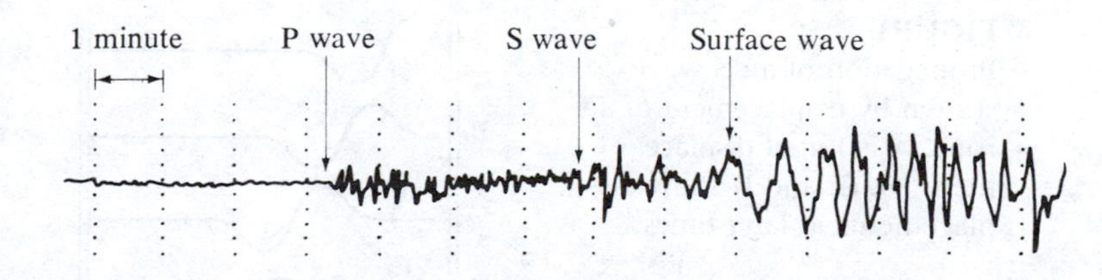

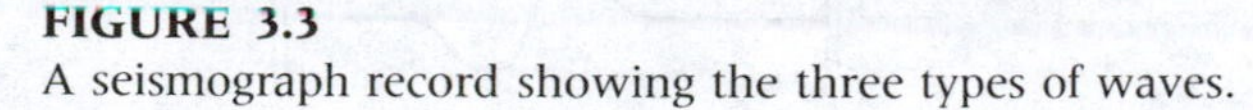

FIGURE 3.3
A seismograph record showing the three types of waves.

FIGURE 3.4
Types of waves. A. A spring stretched between two supports. B. A push wave showing compression and rarefaction. C. A shake wave.

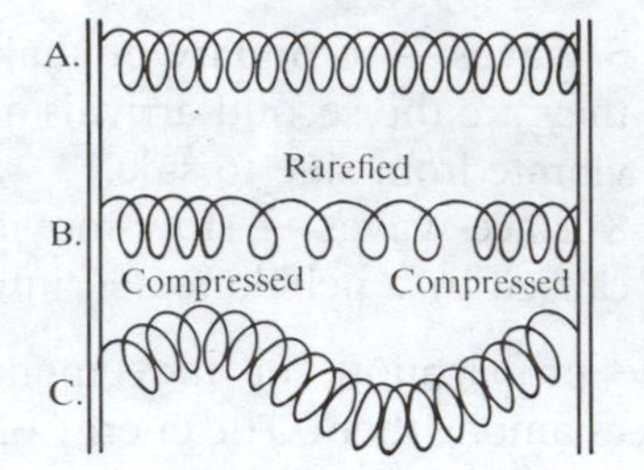

wave moves along the spring by successive displacements to one side and the other (Fig. 3.4). This same motion can be demonstrated by shaking a rope (Fig. 3.5).

The P and S waves from an earthquake vibrate rapidly but with small movement, thus causing very little damage to structures. *The surface waves have much larger amplitudes (movement) and lower frequencies and cause most of the direct damage done by earthquakes.* This is easy to understand, for if one part of a building is moved several inches in one direction while another part moves in a different direction, the building will be damaged. Most earthquake-resistant buildings are designed to be flexible so that such movements will not damage them. The building site will affect the amount of movement that a structure experiences. Thus, buildings on bedrock will be damaged less than those built on less consolidated, easily deformed material such as natural or artificial fills. (These statements apply only to earthquake-wave damage, and do not include the sometimes total destruction that occurs where there is an actual break or displacement at the surface.) Much damage also can be caused by landslides triggered by the earthquake.

Earthquake prediction is one goal of present research. Earthquakes are localized at certain places (a topic discussed later), so that it is possible to prepare earthquake-risk maps such as the one shown in Figure 3.6. Some earthquakes are preceded in certain areas—for example, Japan—by swarms of micro-earthquakes, detectable only on seismographs. At many places, tilting of the ground prior to an earthquake can be detected on instruments. This tilting and associated swelling or rising of the ground is probably caused by many tiny cracks in the deep rocks. These cracks may be caused by the buildup of forces that eventually cause the earthquake.

FIGURE 3.5
Propagation of an S wave shown by displacement of a rope. A is initial displacement; B, C, and D show displacements at later times.

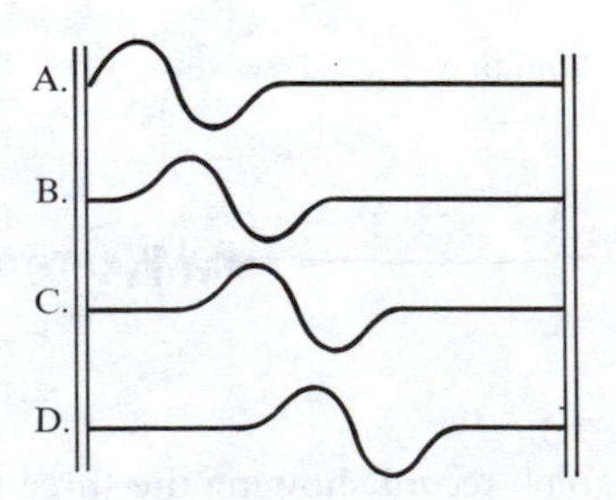

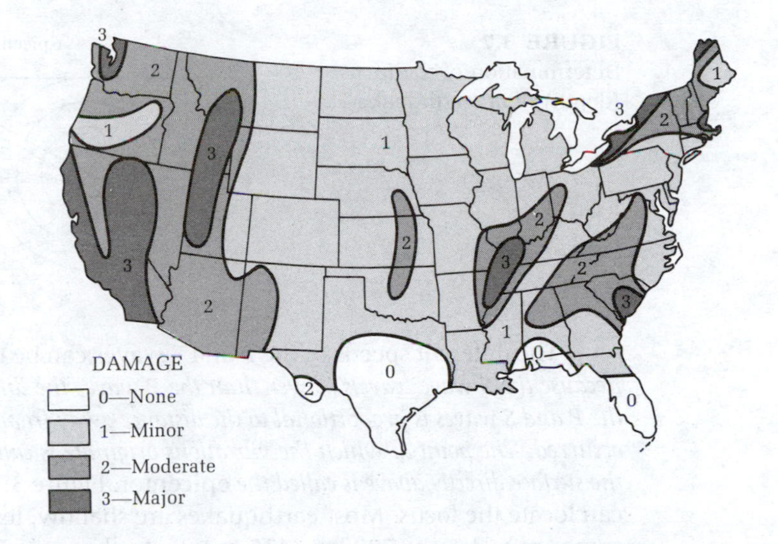

FIGURE 3.6
Seismic risk map for the conterminous United States. The map divides the country into four zones: Zone 0, areas with no reasonable expectancy of earthquake damage; Zone 1, expected minor damage; Zone 2, expected moderate damage; and Zone 3, where major destructive earthquakes may occur. (After Environmental Science Services Administration, 1969)

Another promising method of prediction is the local slowing of P waves from other earthquakes, apparently due to deformation prior to a local earthquake. Use of this method has successfully predicted some earthquakes. Recent studies also have revealed that parts of the San Andreas fault system in California are creeping at rates measured in centimeters per year.

The size of an earthquake can be measured in many ways. The amount of displacement on a fault plane is a measure of near-surface faulting, but the surface break may be different from the bedrock movement at depth because the conditions at the surface may be different from those at depth. Several different scales are in use for comparison of earthquakes. Though the surface effects depend to some extent on the local geology, some scales have been devised that are based on such subjective criteria as noticeable movement and amount of damage. Such reports by witnesses can serve to locate, in a general way, the epicenter of an earthquake. News media report the **magnitude** *of earthquakes, which is a measure of the amount of energy released by the quake.* This measurement is based on the actual movement recorded on a standard type of *seismometer*. The largest magnitude yet recorded on this scale is 8.9, and the smallest felt by humans is 2. The San Francisco earthquake of 1906 was 8.3, and the Loma Prieta earthquake of 1989 was 7.1. For each increase of one unit on the scale—for example, 6.5 to 7.5—the energy released is about 31.5 times greater, and a magnitude 8 earthquake releases about a million times more energy than one of magnitude 4. The atomic bomb detonated over Hiroshima released about as much energy as a magnitude 6.35 earthquake.

Earthquakes that cause damage great enough to be reported in news media are just a few of the total that occur. Although the average annual number of earthquakes strong enough to be felt (at least locally) is estimated at about a million, all but about 1500 are shallow and release little energy.

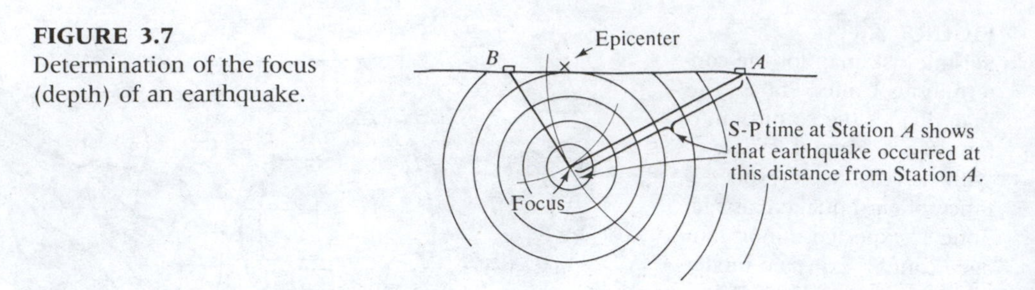

FIGURE 3.7
Determination of the focus (depth) of an earthquake.

The different speeds of the P and S waves can be used to locate an earthquake. *Because the S wave travels slower than the P wave, the difference in arrival time between the P and S waves is proportional to the distance away from the station that the earthquake occurred. The point at which the vibrations originate is called the* **focus,** *and the point on the surface directly above is called the* **epicenter.** Figure 3.7 shows how nearby stations can locate the focus. Most earthquakes are shallow, less than 56 km (35 miles) but some are as deep as 700 km (435 miles). As shown in Figure 3.8, reports from three stations are needed to locate an epicenter.

Travel times of earthquake waves have revealed a layered structure of the earth. The details of the analysis are best seen in Figure 3.9. Three main layers have been distinguished: the **crust,** *a very thin shell between 5 and 56 km (3 and 35 miles) thick;* the **mantle,** *the next layer, about 2900 km (1800 miles) thick;* and *the* **core,** *the central part, with a radius of about 3416 km (2123 miles).* The velocity of earthquake waves changes abruptly at the contacts between these layers, and the S waves cannot penetrate the core. This last observation shows that the core is liquid because shake or shear waves cannot pass through a liquid, although push waves can. Actually, the structure in each layer is probably more complex than suggested here. An example is that P waves speed up as they travel through the central part of the core, suggesting that this part of the core may be solid.

The crust is the thin shell that overlies the mantle. *A very marked change in seismic velocity separates the crust from the mantle. This is the* **Mohorovičić discontinuity,** *commonly called* **M-discontinuity** or **Moho.** It was named for the seismologist who discovered it in 1909. The thickness of the crust (or distance to the Moho) varies between about 5 and 56 km (Fig. 3.11). It is thicker under the

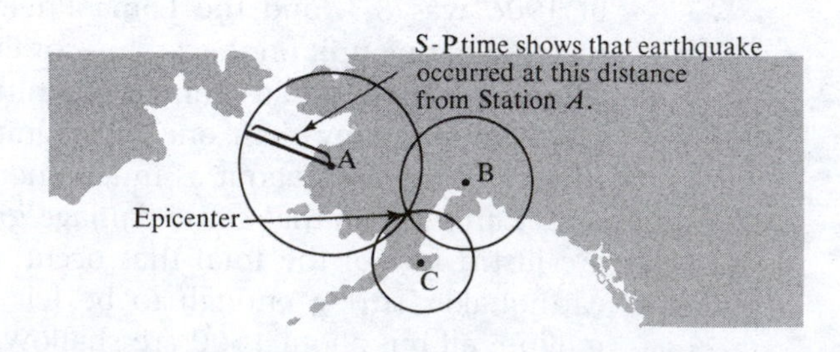

FIGURE 3.8
A, B, and C are seismic stations. The S– P times at each station indicate that the earthquake occurred somewhere on the circle drawn around each station. The point where all three circles intersect is the epicenter.

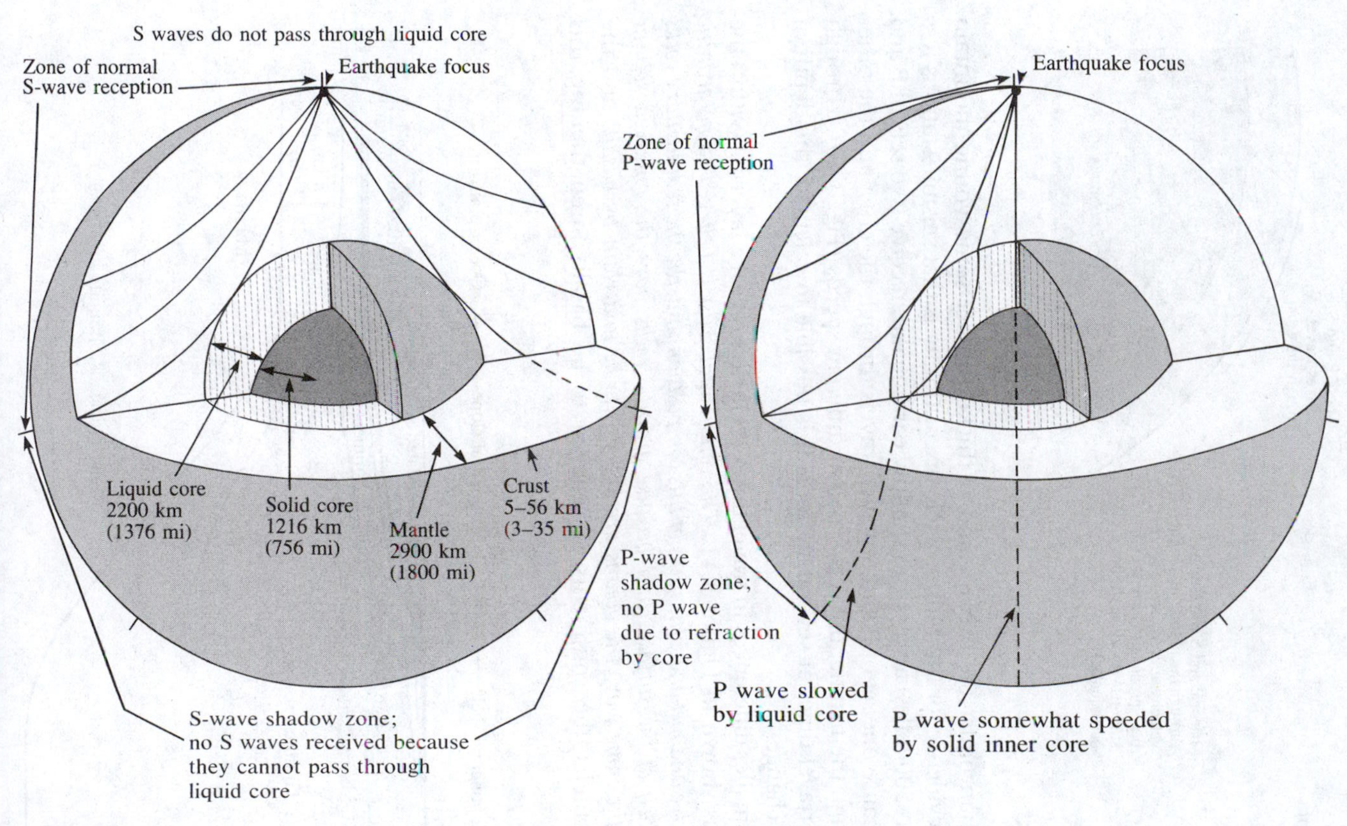

FIGURE 3.9
Behavior of P and S waves in the mantle and the core reveals the layered structure of the earth. Earth is drawn to scale, although crust thickness is exaggerated by the line. Left, behavior of S waves; right, behavior of P waves. In the earth the travel paths are curved because of the increase in velocity with depth in each layer.

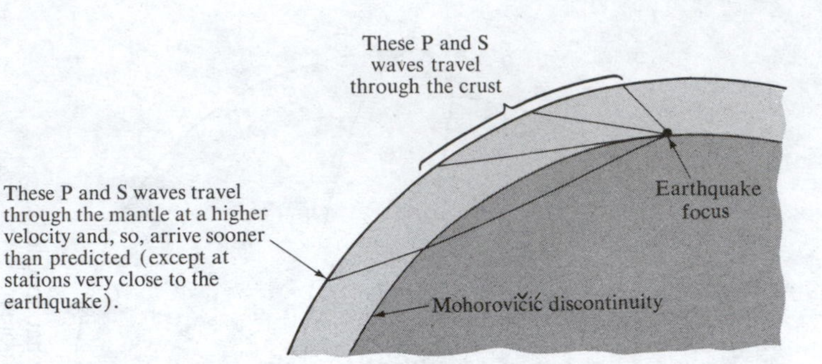

FIGURE 3.10
Determination of the depth of the Mohorovičić discontinuity.

continents than under the oceans, and is thickest under some young mountain ranges. The evidence for the Mohorovičić discontinuity is that seismic stations less than 800 km (500 miles) away from shallow earthquakes receive two sets of P and S waves. Because one set of P and S waves travels through the crust and the other mainly through the mantle, the sets travel at different speeds (Fig. 3.10). Through study of the travel times at several stations, it is possible to calculate the depth of this velocity change.

Other smaller differences in travel times (velocity) reveal a layering within the crust itself, as shown in Figure 3.11. *The continental crust is believed to be granitic because intrusive and metamorphic rocks of this composition underlie the sedimentary rocks that cover much of the surface, and because granitic rocks have the observed seismic velocities. In the same way, the oceans are believed to be underlain by basalt,* and the lower layer under the continents is believed to be basalt. Strengthening this

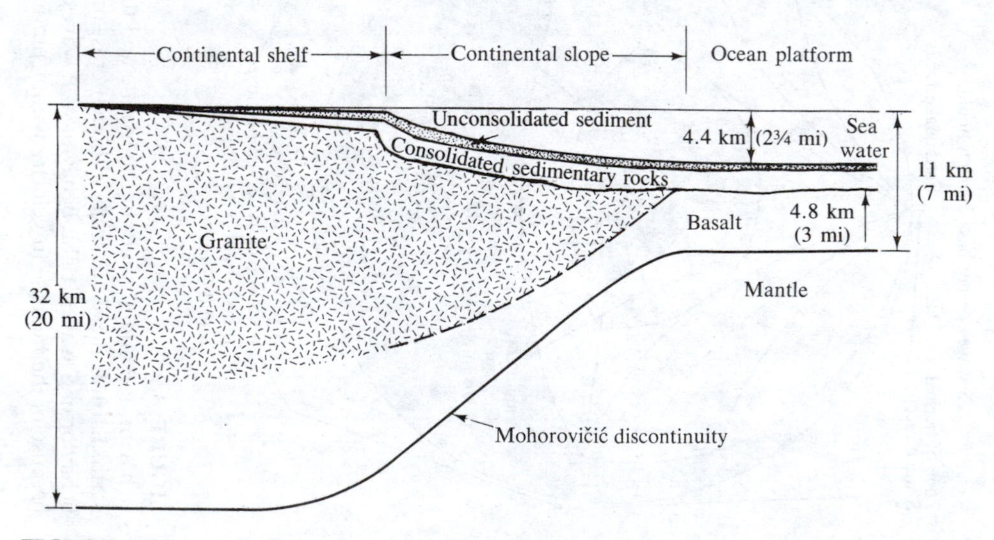

FIGURE 3.11

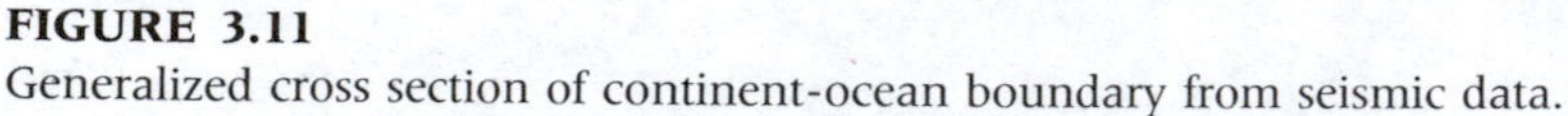
Generalized cross section of continent-ocean boundary from seismic data.

inference is the fact that basalt volcanoes occur in both oceanic and continental areas. (Basalt is used here to designate composition only; the rocks may be crystallized as gabbro or even amphibolite.)

The nature of the Mohorovičić discontinuity is not known, except that there is a marked change in seismic velocity between the mantle and the crust. For many years this was assumed to be caused by a change in composition from the peridotite or dunite of the mantle to the basalt of the lower layer of the crust. Recently, however, it has been suggested that the change may not be due to a difference in chemical composition, as between dunite and basalt, but to a change in mineralogy in a rock of the same chemical composition. Such changes are known to take place as a result of increased pressure and might occur in a zone narrow enough to account for the rapid change in velocity. This type of change occurs when basalt is subjected to high temperature and pressure. Under these conditions, the augite and plagioclase of basalt become unstable, and the elements recombine to form a type of garnet and a green pyroxene, omphacite. Such rocks are called *eclogite,* and their rare appearance at the surface is at places where uplift and erosion have exposed deep-seated rocks or where intrusive bodies have carried them to the surface. The weight of the overlying rocks supplies the pressure needed to change the mineralogy. Therefore, if the Moho is in fact caused by changes in mineralogy rather than in composition, it should follow the surface topography; however, it does not.

Although the Mohorovičić discontinuity that separates the crust from the mantle is a marked seismic velocity change, a much less easily detected *low-velocity zone* in the upper mantle may be much more important in understanding the structures of the crust. This low-velocity zone is discussed in later sections of this chapter.

The velocity of waves is related to the physical properties of the medium through which they pass; hence, it is possible to make some inferences about the material that composes the deep parts of the earth. Remember also the density distribution of the deeper parts of the earth that was learned from the earlier discussion of the earth-moon system. Both methods of investigation lead to the specific gravity of the mantle varying from about 3 near the top to about 6 at its base. Similarly, the core has a specific gravity of about 9 at its top and about 12 at the center. These inferences are combined with our knowledge of meteorites, which are believed to be of material similar to that from which the earth was formed, to develop a model of the earth.

Composition of the Layers of the Earth

Meteorites *are objects that fall onto the earth from space.* Many of them are burned by frictional heat in the atmosphere to form what are called "shooting stars." Some meteorites are quite large and form craters, such as Meteor Crater in Arizona. Many of the craters on the moon, which is not protected by an atmosphere, are believed to have formed by meteorite impact.

Meteorites are of three types: *stony* meteorites, composed mainly of olivine and pyroxene (peridotite is the rock name for this composition); *iron* meteorites, composed mainly of iron and nickel; and *stony-iron* meteorites, which are mixtures.

The specific gravity and other physical properties of peridotite or dunite fit those inferred for the mantle. In the same way, the core is believed to be molten iron and nickel. In reaching these assumptions, allowance must be made for the increased temperature and pressure deep within the earth, for these will greatly modify the properties of rocks within the earth. These temperatures and pressures are too high to reproduce in the laboratory, so some uncertainties exist.

The pressure deep in the earth is reasonably well known, but the temperature is less so. The pressures are due to the weight of the overlying material, and, with our knowledge of specific gravity gained from seismic studies, can be calculated from the law of gravity (Fig. 3.12).

The temperatures in the earth are estimated from the physical properties of the rocks because the origin of the heat is not known. The earth's heat may be a remnant of its formation, or may be due to radioactive sources in the earth. (It clearly is not due to heating by the sun, because heat from the sun can be accounted for, and it only maintains the surface temperature.) Data from drill holes and deep mines reveal that the temperature increases with depth in the earth at the approximate rate of 1°C per 28 meters. If this rise continues to the center of the earth, the temperature there would be 225,000°C, an unreasonably high temperature—hotter than the sun.

The upper limit of temperatures in the earth is estimated from knowledge of the melting points of the materials assumed to form each layer, taking into account the rise in melting point at high pressures. The lower limit is estimated from calculations of the lowest temperature that will permit heat transfer by convection. Convection is believed to be an important process in the mantle and is discussed later. Although convection requires a fluid, the material in the mantle could flow slowly (and still transmit S waves as a solid), in much the same way that tar

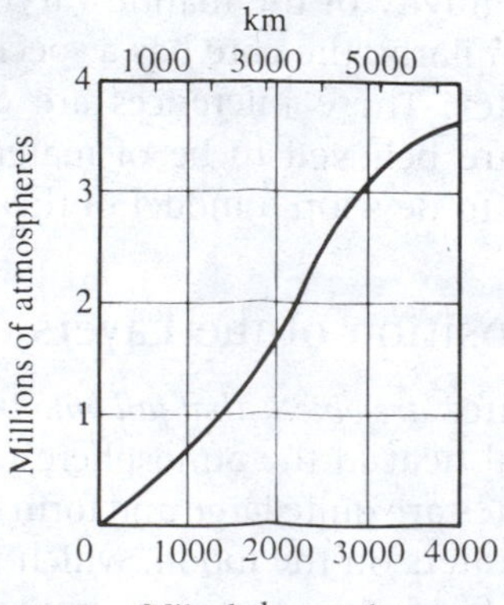

FIGURE 3.12
Pressures within the earth. (Adapted from John Verhoogen, *American Scientist*, Vol. 48, 1960)

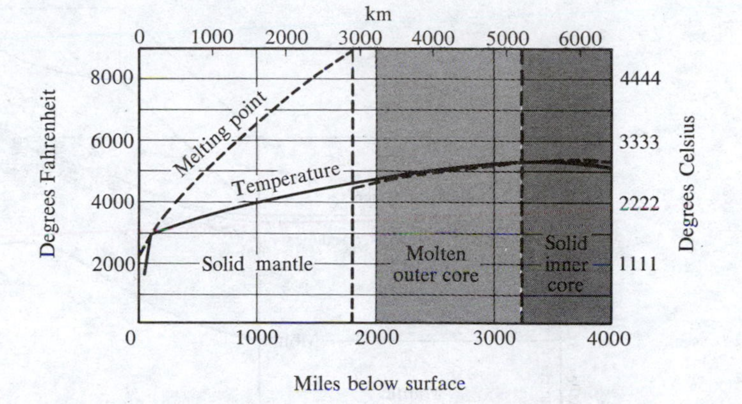

FIGURE 3.13
Estimated temperatures in the earth. (Adapted from John Verhoogen, *American Scientist*, Vol. 48, 1960)

behaves at room temperature. Under such conditions, tar acts as a solid to rapid deformation, such as the passage of S waves; but over a long period, it will flow under its own weight. From this reasoning, the temperature distribution shown in Figure 3.13 is derived.

STRUCTURE OF THE CRUST

Structure of the Oceans

Ocean water covers more than 70 percent of the earth's surface. Thus, except for a few bottom samples, our knowledge of the ocean bottom and most of the border areas comes mainly from indirect means such as depth soundings, seismic studies, and some drilling. As a result, only large-relief features are well known.

Ocean Basin Floors and Submarine Mountains The **ocean basin floors** are relatively smooth surfaces at an average depth of about 4420 m (14,500 feet) that cover about 40 percent of the earth's surface. Seismic studies reveal that, on the average, the ocean floors are covered by 300 m (1000 feet) of unconsolidated sediments and, below these, by a few thousand meters of presumed consolidated volcanic and sedimentary rocks that overlie about 4900 m (16,000 feet) of crustal material, probably basaltic, above the Mohorovičić discontinuity (Fig. 3.14). The age and distribution of these rocks are very important and will be described later.

Extensive mapping of the oceans with echo sounders (Fig. 3.15) was started just before World War II and continued during and after the war. Much was soon learned about submarine mountain ranges, and mid-ocean ridges were discovered.

The **submarine mountains** *are basalt volcanoes.* Some, such as the Hawaiian Islands, extend above sea level. In some areas these mountains are flat topped. These flattops are now at depths up to about a mile below the surface of the ocean. They are especially interesting because their flat tops were probably produced by

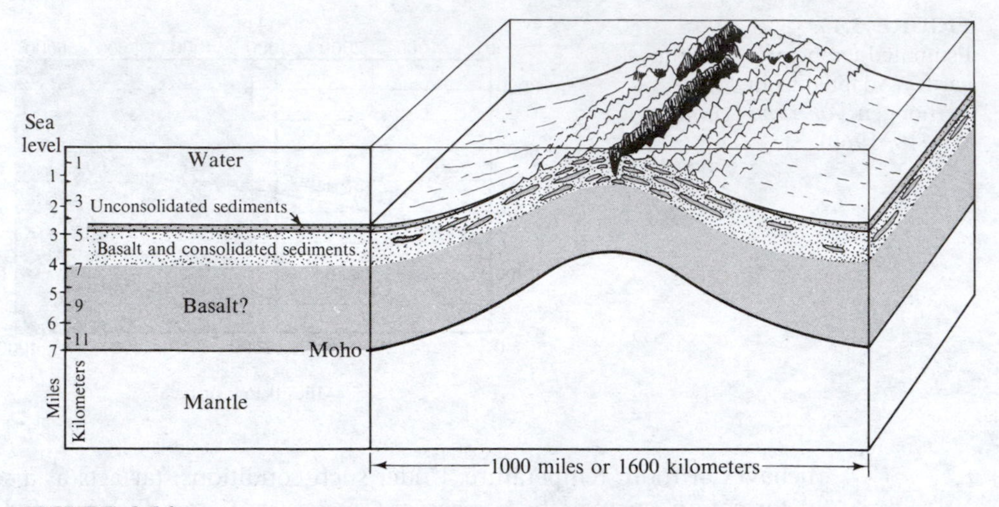

FIGURE 3.14
Mid-ocean ridge. A cross section of typical ocean-basin floor is shown on the left. The ocean crust thins at the ridge. Note the vertical exaggeration.

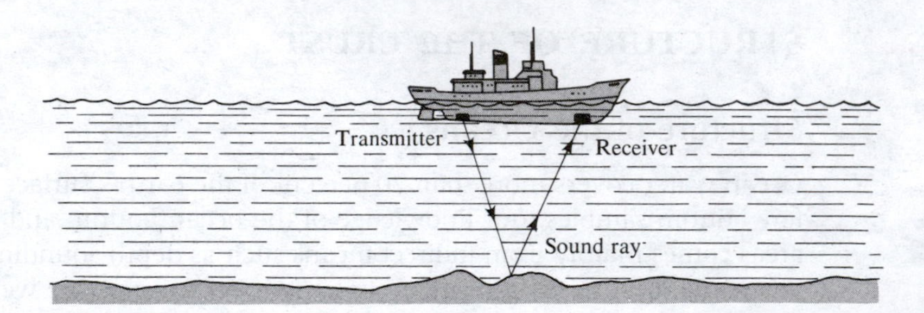

FIGURE 3.15
An echo sounder. Depth = 1/2 speed of sound × echo time.

wave erosion at sea level. Several lines of evidence point to this conclusion. Wave erosion is one of the few ways that such surfaces can be produced, although, under some conditions, small submarine volcanoes may form with initial flat tops. Wave-cut benches are being formed now on the present islands, and if this process continues, such islands will become flat topped. Lowering of sea level or rising in elevation of an island can preserve such benches, and this too can be seen on present-day islands and coastlines. Similar benches are revealed in profiles of some of the flat-topped seamounts (Fig. 3.16). Samples dredged from flat-topped seamounts contain basalt fragments and Cretaceous and early Cenozoic shallow-water fossils, suggesting the age of the wave erosion.

The *coral* **atolls** of the Pacific also may reveal subsidence. Corals live in the near-surface zone and form an offshore reef (Fig. 3.17). If the island sinks, the

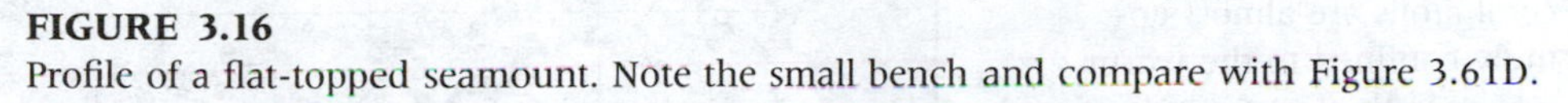

FIGURE 3.16
Profile of a flat-topped seamount. Note the small bench and compare with Figure 3.61D.

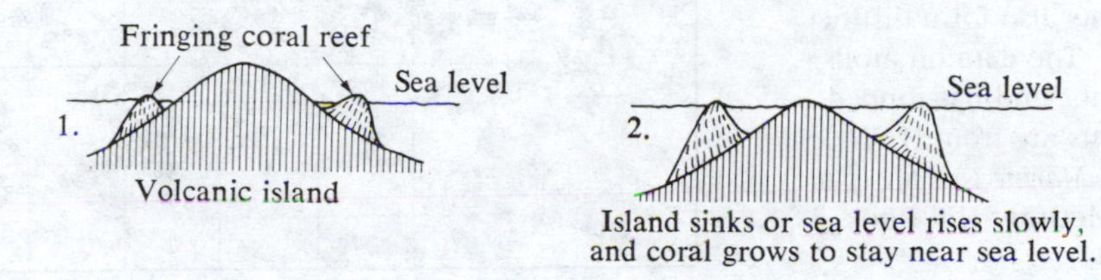

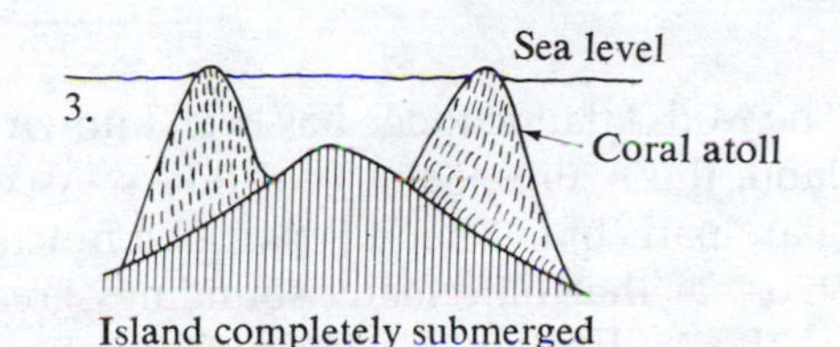

FIGURE 3.17
Development of a coral atoll.

corals will build up the reef so that they remain near the surface where their food supply is. This process may continue until the island is completely submerged, perhaps quite deeply, and only a coral reef remains. At Eniwetok Atoll in the western Pacific, the coral is 1370 m (4500 feet) deep. Thus, the flat-topped seamounts tell a story of volcanic activity, wave erosion, and rapid subsidence; if they had subsided slowly, coral would have been able to build up and form an atoll. Figure 3.18 shows the distribution of seamounts and associated features in the Pacific Ocean.

Mid-Ocean Ridges The **mid-ocean ridges** *form the longest continuous mountain range on the earth.* Their extent can be seen on the map in Figure 3.29. Their length is about 64,000 km (40,000 miles); their width is 480–4800 km (300–3000 miles); they rise to about 3000 m (10,000 feet) above their base; and at places the highest points form islands. Parts of the mid-ocean ridge go inland in Africa, and the high plateaus and rift valleys there may be associated with it. Two parts of this ridge are fairly well known: the Mid-Atlantic Ridge and the East Pacific Rise.

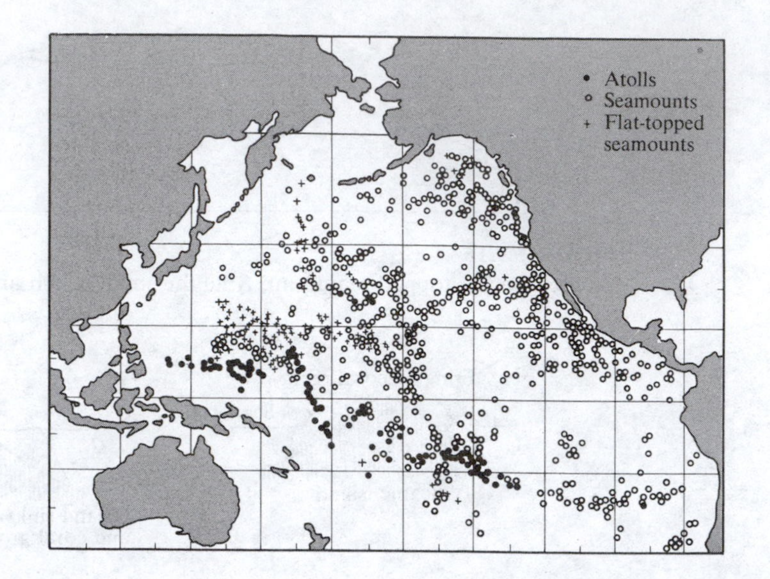

FIGURE 3.18
Map of the Pacific Ocean showing atolls, seamounts, and flat-topped seamounts. These volcanic features tend to form lines, suggesting eruption along faults. The coral atolls are almost entirely confined to the warm waters within 30 degrees of the equator. The flat-topped seamounts also form distinct clusters. (The data on atolls, seamounts, and flat-topped seamounts are from H. W. Menard, *Marine Geology of the Pacific,* McGraw-Hill Book Co., 1964)

The Mid-Atlantic Ridge has been known for many years because islands such as Iceland, the Azores, Saint Paul Rocks, Ascension, Saint Helena, and Tristan da Cunha are parts of it. That it is part of a much larger feature was not known until the 1950s; at that time many soundings greatly increased our knowledge of its nature. The Mid-Atlantic Ridge consists of numerous parallel ridges and lines of peaks that extend for hundreds of kilometers on the flanks (Fig. 3.14). The central crestal zone has one or more discontinuous rift valleys that are up to 48 km (30 miles) wide. This topography suggests that the ridge was formed by a series of uplifted fault blocks. The bedrock, revealed in dredged samples, is basalt, and some of the peaks that form in lines are apparently basalt volcanoes that formed along the faults. At many places these features are apparently offset large distances, as can be seen in Figure 3.29. The unexpected nature of the movement on these faults, or fracture zones, is described later. Many earthquake epicenters are located on the mid-ocean ridges.

The East Pacific Rise, a part of the mid-ocean ridge, forms the Galápagos and Easter Islands. It meets North America just south of Baja California, leaves in northern California, and is last seen off British Columbia. The San Andreas Fault in California may be associated with the East Pacific Rise. Measurements on the East Pacific Rise show that the heat flow near the crest is up to five times greater than average, and on the flanks is lower than average. Because the rocks under the ridge are hot and therefore less dense, this may be one reason the ridge is higher than the surrounding ocean floor. Seismic studies using underwater explosions have revealed that the crust under the East Pacific Rise is only about three-quarters as thick as normal ocean crust.

Structure of the Continent-Ocean Border

Continental Slope The borders between the oceans and the continents are of at least two types. The simpler of these is the **continental slope** shown in Figure 3.19. This type of border occurs in more geologically stable areas than the volcanic arcs described below. The east coast of the United States is generally this type of border. The main features are the **continental shelf,** *which has continental structure but is submerged, and the continental slope itself, which is the actual border.* The continental slope is a few tens of kilometers wide, and seismic studies show that it is the transition where the granitic continental crust thins very rapidly and the deep ocean begins. The continental shelves are generally covered with sedimentary rocks that thicken on the continental slopes.

Volcanic Arcs At other places, particularly on the rim of the Pacific Ocean, the border is more active, with chains of volcanic islands forming arcs (Fig. 3.20). The **volcanic arcs** were noted early in the history of geologic oceanography, but recognition of many of their features came much later. Although oceanography began in the 1870s with the voyage of the *Challenger,* little was learned in detail until the 1920s when echo sounders were first employed. It was soon discovered that deep trenches exist on the ocean side of the volcanic island chains. During the 1920s and early 1930s, Dutch geologist F. A. Vening-Meinesz devised a method of measuring the value of gravitational attraction at sea. Using a method that measured the period of a pendulum, he had to work in Dutch and U.S. Navy submarines to avoid the effects of ocean waves. He made the extremely important discovery that the *acceleration of gravity is low* **(negative gravity anomaly)** over these deep trenches. This discovery shows that there is less mass directly below the trench (Fig. 3.21). The volcanic arcs also are active seismically, and about the same time it was discovered that some of the earthquakes come from depths up to 700 km (435 miles). These deep-focus earthquakes occur mainly on the Pacific rim. Their foci lie on a plane continuous with the shallow and intermediate earthquakes (Figs.

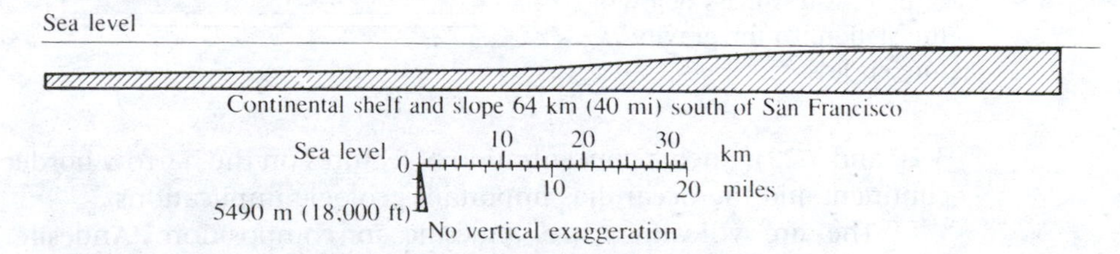

FIGURE 3.19
Continental shelf and slope with no vertical exaggeration. Compare with Figure 3.11, which shows the structure. (After K. O. Emery and others, U.S. Geological Survey Professional Paper 260 A, 1954)

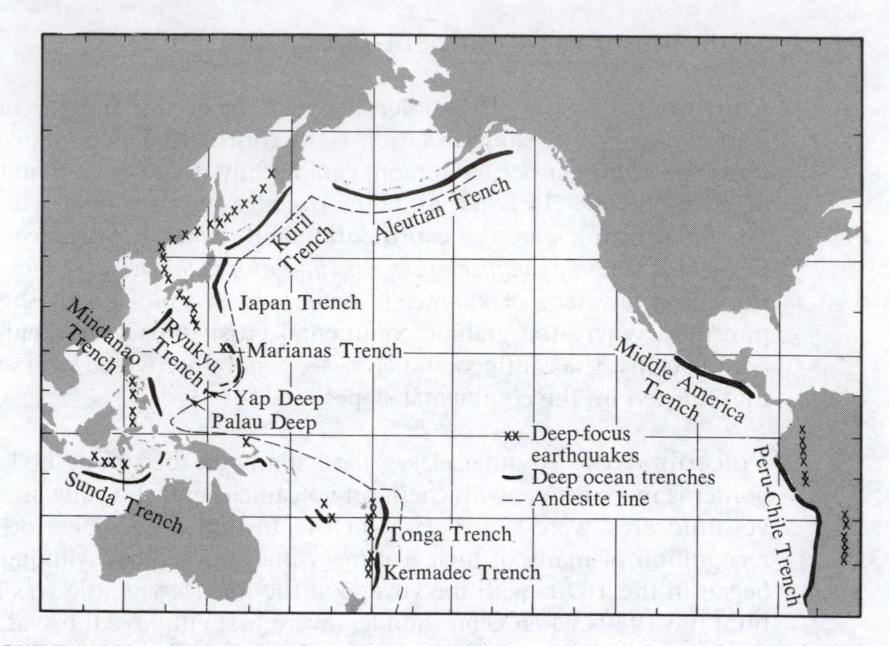

FIGURE 3.20
Map of the Pacific Ocean showing the relationships among volcanic arcs, deep-focus earthquakes, and deep trenches. The volcanic rocks on the continent side of the andesite line are andesite, and basalt occurs on the ocean side. The epicenters of shallower earthquakes lie between the deep earthquakes and the volcanic arcs.

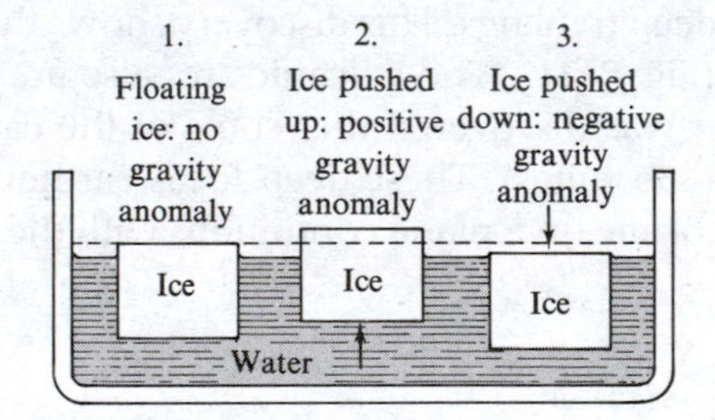

FIGURE 3.21
How gravity anomalies are formed: Station 1 gravity is normal. At Station 2, the attraction of gravity is higher because more mass is present below the station. At Station 3, there is less mass below the station, so the gravity force is less.

3.20 and 3.22). Such a concentration of features on the narrow border between the continent and the ocean has important geologic implications.

The arc volcanoes are andesitic in composition. Andesite also is the composition of volcanoes in the Andes Mountains of South America and the Cascade Range in North America, which, together with the island arcs, form what has been somewhat poetically called the "rim of fire" of the Pacific Ocean. The deep oceanic trough off South America and the deep-focus earthquakes inland suggest that the structure there may be similar to the volcanic island arcs of the western

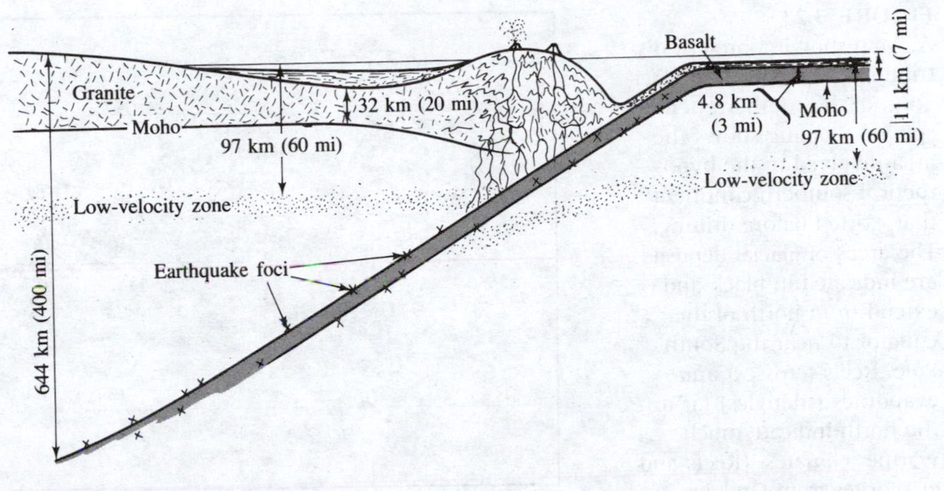

FIGURE 3.22
Volcanic island arc showing location of earthquake foci. Many of the structural features shown here are discussed later in this chapter. (Because of the distances involved, this figure is not drawn to scale.)

Pacific. On the oceanward side of this rim of andesite volcanoes, all of the volcanic islands are basaltic (see andesite line in Fig. 3.20).

These oceanic features are involved in shaping the earth's surface, but some other aspects must be considered before we return to the role of the oceans.

Continental Drift

The idea that the continents have moved **(continental drift)** has been with us since the first accurate maps showed that Africa and South America fit together. During the last part of the nineteenth century, geologists working in India, Australia, and South Africa discovered ancient glacial deposits and, associated with them, distinctive fossils, especially of plants. Similar discoveries were made later in South America and Antarctica. As shown in Figure 3.23, these glacial deposits are scattered from north of the equator to nearly the south pole. When these discoveries were made, many geologists working in the southern hemisphere advocated continental drift to explain these features, even though no mechanism to cause it was known. Recently, more evidence favoring continental drift has been discovered.

Earth's Magnetic Field

Rock magnetism is the most important new evidence for continental drift. The earth's magnetic field is impressed on a rock at the time the rock is formed. Basalt, because it contains small amounts of magnetite, an especially magnetic mineral,

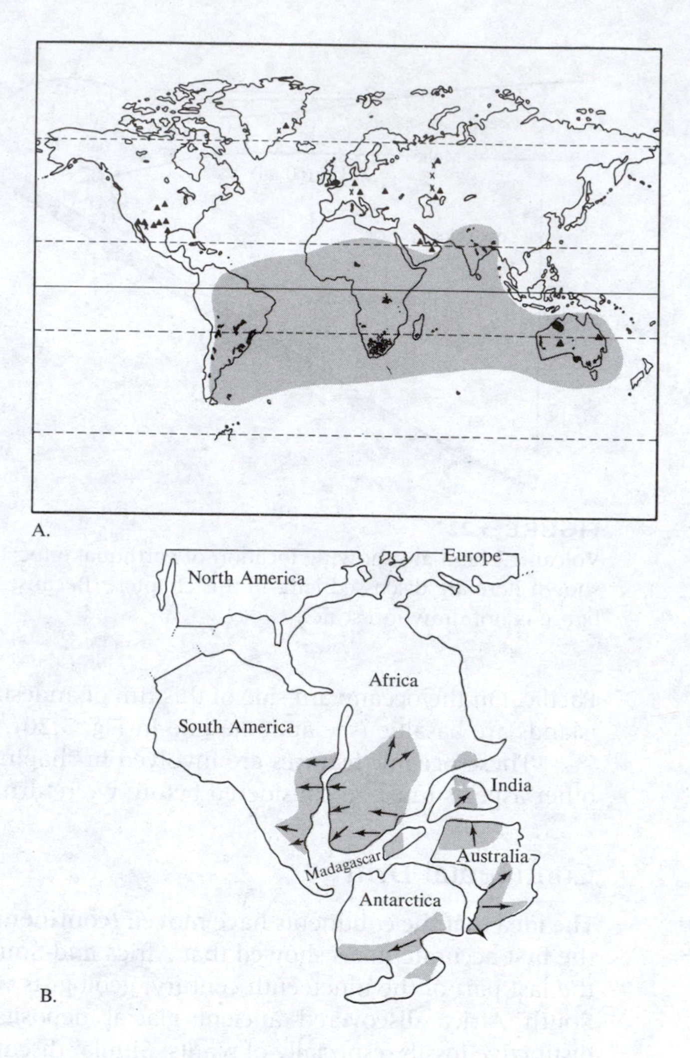

FIGURE 3.23
A. Map showing present distribution of the ancient glaciation of the southern hemisphere. Shading shows the area involved in the hypothetical southern continent that existed before drifting. The areas of glacial deposits are indicated in black and extend from north of the equator to near the South Pole. Reefs (crosses) and evaporites (triangles) far to the north indicate much warmer climates. (Reefs and evaporites from Opdyke, 1962) B. Reconstruction of the ancient continent of the southern hemisphere. The shading and arrows show the glaciers and their movement. Dotted lines indicate changes in shapes of continents.

faithfully records the earth's magnetic field at the time the basalt cools through a certain temperature that is well below its melting point.

To understand rock magnetism, a closer look at the nature of the earth's magnetic field is necessary. A magnetic compass points toward the earth's magnetic north pole, which is about 11.5 degrees away from the geographic North Pole, because the compass needle aligns with the earth's magnetic field. If, however, the compass is free to pivot in the vertical plane as well as the horizontal, it will align parallel with the earth's field and point downward at an angle as well as towards north. This is shown in Figure 3.24, where it can be seen that the vertical angle depends on the latitude.

Thus, the magnetic field recorded in a rock, such as basalt, shows both the latitude and the direction to the magnetic north pole at the time the rock formed, assuming, of

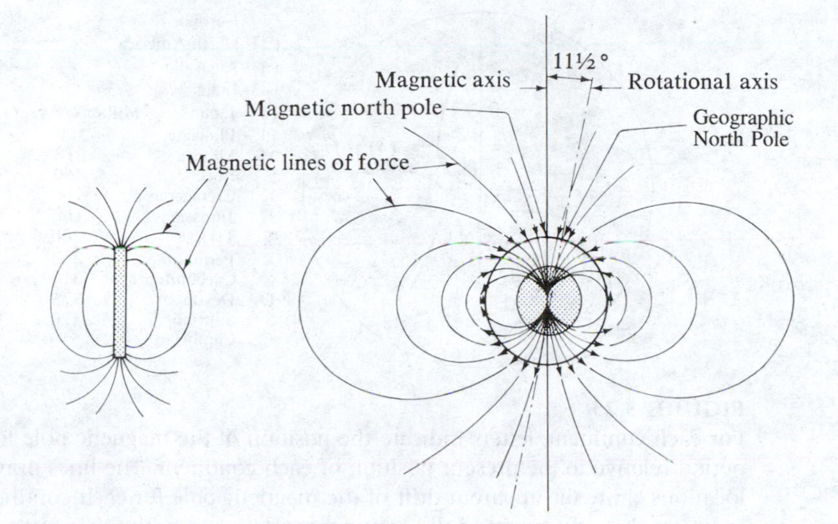

FIGURE 3.24
The earth's magnetic field is much like that of a bar magnet. The magnetic poles do not coincide with the geographic poles, which are determined by the rotation axis. The short, heavy arrows show the vertical direction in which a compass will point if free to move in the vertical plane.

course, that the earth's magnetic field then was similar to the present field. This latter assumption is made in all studies of fossil magnetism and may be a source of difficulty, because the origin of the earth's field is not known with certainty. The earth's field is believed to be caused by electric currents in the iron-nickel core, and all features of the earth's magnetism are consistent with this theory.

Measurements of the direction and angle of the magnetic field of rocks of all ages have been made on all continents. From each measurement the position of the magnetic north pole has been calculated. When these positions are plotted on a map, it is clearly seen that the poles for any one age are at a different position for each continent, strongly suggesting movement of the continents relative to each other (Fig. 3.25). Because magnetic data can show only latitude and north direction, it is not possible to locate the place on that latitude where the continent was, as shown in Figure 3.26. For this reason, it is not possible to reconstruct actual positions of the continents in the past, or their migration routes, from magnetic data alone.

To find actual positions of the continents and their migration routes, it is necessary to make a number of assumptions and to match rocks like pieces of a jigsaw puzzle. The puzzle, however, is a multilayered one, and erosion and deformation have destroyed or altered many of the pieces.

In summary, continental drift is suggested by the paleomagnetic data just described, the fit of the continents, the matching of rocks and structures, similarity

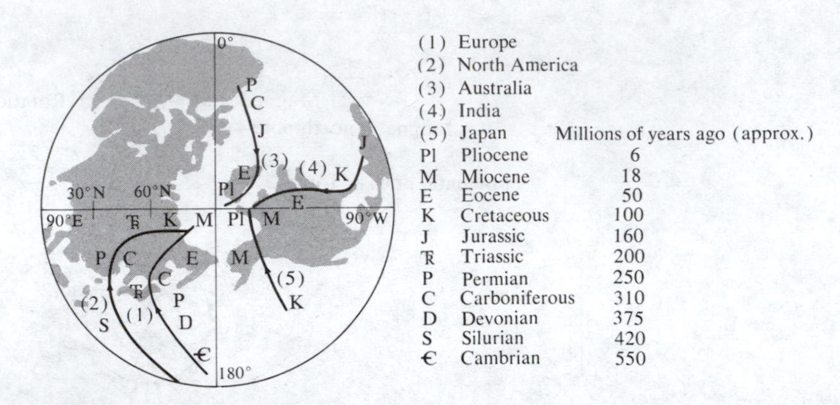

FIGURE 3.25
For each continent, letters indicate the position of the magnetic pole for each geologic period relative to the present position of each continent. The lines drawn through these locations show the apparent drift of the magnetic pole for each continent. The diagram indicates that movement of the continents relative to each other must have occurred but does not show the route of drifting because of the limitations of paleomagnetic data shown in Figure 3.26. (After Allen Cox and R. R. Doell, *Geological Society of America Bulletin,* Vol. 71, 1960)

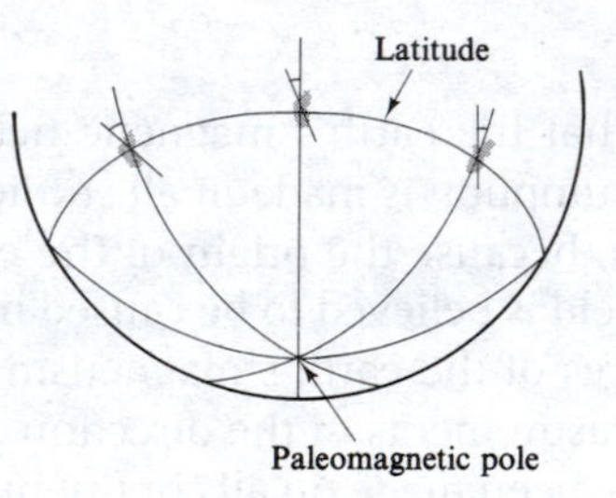

FIGURE 3.26
Paleomagnetic data can show only the latitude and orientation of a continent. The horizontal component of the paleomagnetism points the direction toward the pole, and the vertical inclination gives the latitude. The continent could have been in that orientation anywhere on that latitude.

of fossils, and similarity of climate shown by glacial deposits. Evidence against drift is given by possible changes in the nature of the earth's magnetic field in the past and the similarity in heat flow from continental and oceanic crust. This latter point is that the heat flow measured from the continents is heat flow due to the radioactive elements in the continental crust, but the oceanic crust contains much less radioactive material so its heat flow must come largely from the mantle. This suggests that the mantle under the oceans is different from that under the continents or, said another way, that the difference between continent and ocean may extend deep into the mantle, perhaps too deep for drift to be possible.

Sea-Floor Spreading—Movement of the Sea Floor

Sea-floor spreading is the theory that oceanic crust is created at mid-ocean ridges and then moves toward the volcanic arcs where it is consumed (Fig. 3.27). The volcanic activity at mid-ocean ridges suggests that new crust could be forming there, and the age of the ocean floors strengthens this idea. The age of the sea floors is determined from geologic study of sediments and the magnetic properties of the oceanic crust. Magnetic evidence was crucial in the establishment of the theory of sea-floor spreading.

The evidence that new sea floor is created at the mid-ocean ridges is abundant. Radioactive dating of basalt samples shows a progressively older age away from the ridges. The sediments on the sea floor also thicken away from the ridges, and examination of drill cores reveals that the bottom sediments are older farther away from the ridges. The sea floors are young geologically, and only a few samples older than 180 million years have been found.

Magnetic Stripes The earth's magnetic field is recorded in the ocean-floor basalt at the time it cools. Magnetic surveys near the mid-ocean ridge show stripes of alternating high and low magnetic intensity that parallel the ridge, as shown in Figure 3.28. The meaning of these magnetic bands was not clear until magnetic studies were made on the continents. It was soon discovered that many rocks recorded a magnetic field opposite or reversed to the present magnetic field. Magnetic studies of each flow in areas of thick accumulations of basalt flows showed that these reversed magnetic field rocks did not occur in a random way, but, rather, that if radioactive dates of the rocks also were obtained, the magnetic reversals occurred simultaneously all over the earth.

Starting with the youngest rocks, it has been possible to determine the times of reversals accurately enough to use the magnetic direction to date some young rocks. These reversals suggested that the bands of alternating high and low magnetic intensities paralleling the mid-ocean ridge could be caused by bands of ocean-floor basalt that have normal and reversed magnetic fields. The normally magnetized basalt would reinforce the present magnetic field, giving a high magnetic intensity; and the reverse-magnetized rocks would partially cancel the

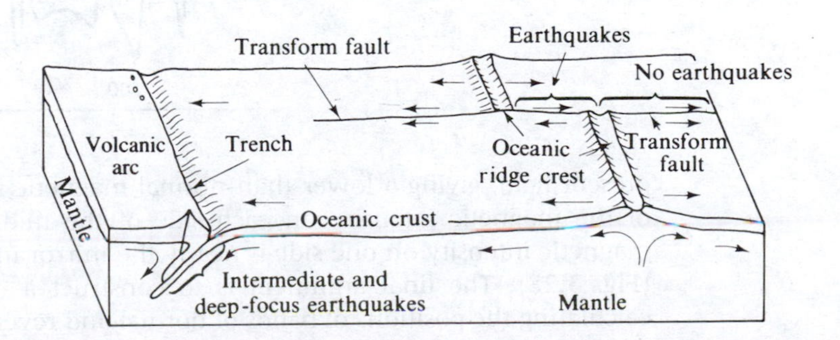

FIGURE 3.27
Sea-floor spreading. Oceanic crust is created at the ridges and consumed at the trenches. Note the location of earthquakes on the transform fault.

FIGURE 3.28
A. Formation of magnetic anomalies at mid-ocean ridges. In this hypothesis, the lavas that form the ocean floor cool at the ridge and are magnetized by the earth's field at the time they cool. The lavas move away from the ridges and so create magnetic anomaly patterns parallel and symmetrical to the ridges. B. The two upper curves are an actual magnetic profile of part of the East Pacific Rise, shown both as recorded and reversed. The lower curve was computed from a model, assuming a spreading rate of 4.4 cm per year. (B. After F. J. Vine, *Science,* Vol. 154, p. 1409, December 16, 1966, © 1966, American Association for the Advancement of Science)

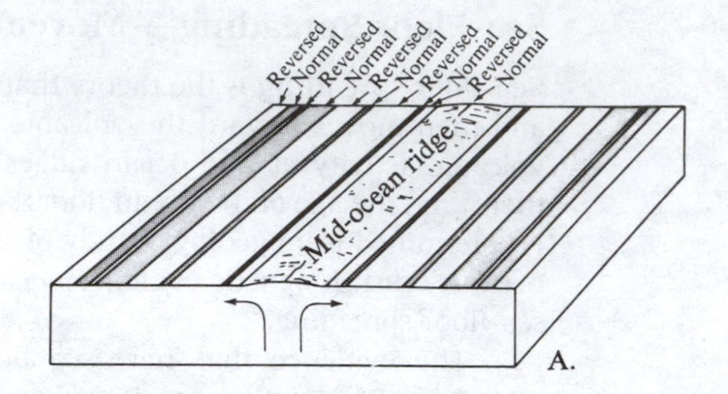

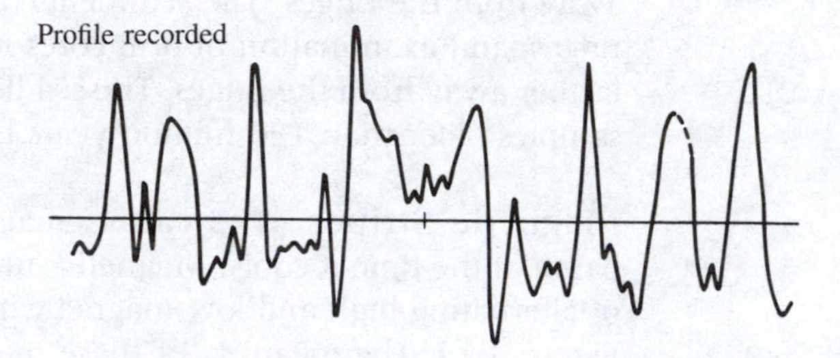

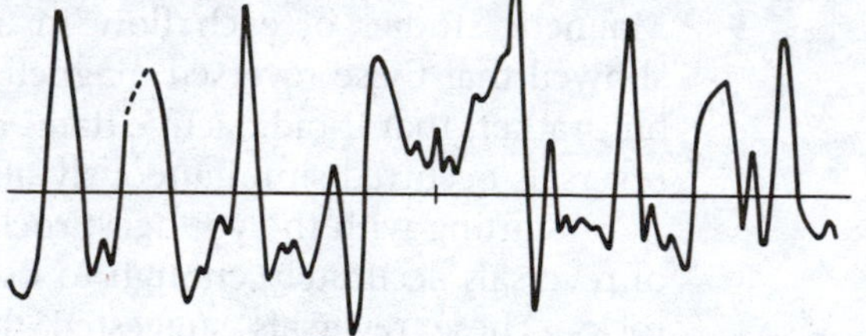

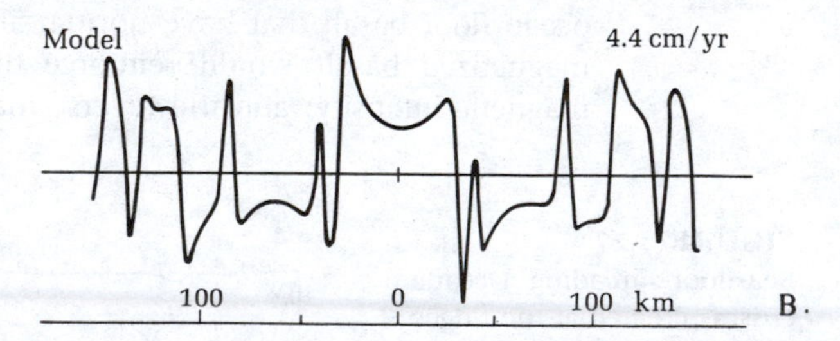

present field, giving a lower-than-normal magnetic intensity. Careful comparison of the magnetic intensity on each side of the mid-ocean ridge shows that the magnetic intensity on one side is nearly the mirror image of that on the other side (Fig. 3.28). The final argument is to construct a model of the ocean floor by calculating the positions of bands of normal and reversed magnetic basalt and then to calculate the theoretical magnetic field produced by such a model. As shown in Figure 3.28, this model produces a field almost identical to that found in nature.

Plate Tectonics

Plate tectonics is the theory that unifies all of the preceding observations. All of the elements of plate tectonics already have been stated. The plates are created at mid-ocean ridges, consumed at volcanic arcs, and carry continents with them as they move. The newly created ocean floor moves away from the mid-ocean ridge at rates generally between 1.3 and 5 cm (1/2 and 2 inches) per year.

The active plates are outlined by earthquakes. Figures 3.27 and 3.29 show where earthquakes occur; their concentration is at the deep-sea trench–volcanic arc areas, to a lesser extent on the mid-ocean ridge, and at many places on the faults or fracture zones that offset the ridge. These earthquakes suggest that the earth can be thought of as consisting of seven main plates and a few smaller ones, each moving from the mid-ocean ridge to a trench or volcanic arc. These plates are indicated in Figure 3.29.

The map shows that plates have three types of borders: the mid-ocean ridges (spreading centers, divergent boundaries) where they are created; the subduction zones (convergent zones) where they are consumed; and transform faults where they move past each other. If the plates are active, earthquake maps show these boundaries clearly; but if a plate is inactive, its margin may be difficult to find.

A plate may be composed of ocean only, or it may carry a continent as well. The plates are rigid, meaning that the distances between points on a plate do not change as the plate moves. Distances between points on different plates do, of course, change.

Transform Faults and Fracture Zones Although the mid-ocean ridge can be followed continuously through all of the oceans, its center is clearly offset at many places (Figs. 3.27 and 3.29). The map pattern suggests that movement along faults has displaced the ridge, and this idea was strengthened by **fracture zones** (to be described), extending many hundreds of kilometers from the offsets. However, this simple interpretation is *not* correct, and the offsets of the mid-ocean ridge or spreading center are more complex. The features that offset the ridge are called *transform faults.*

Mid-ocean ridges probably originate as discontinuous features; that is, the apparent offsets are original features. In any case, the movement on transform faults caused by formation of new ocean crust at segments of mid-ocean ridges is in the opposite direction to the apparent offset (Fig. 3.30). Detailed seismic studies reveal the nature of the movements on the faults that offset the mid-ocean ridge. It is possible to determine the relative direction of the movement of the plates on each side of the fault from the direction of the first movement on seismograms. Such studies show that the movement on each block is away from the mid-ocean ridge, as expected if new crust is created there. Note, however, as shown in Figure 3.30,

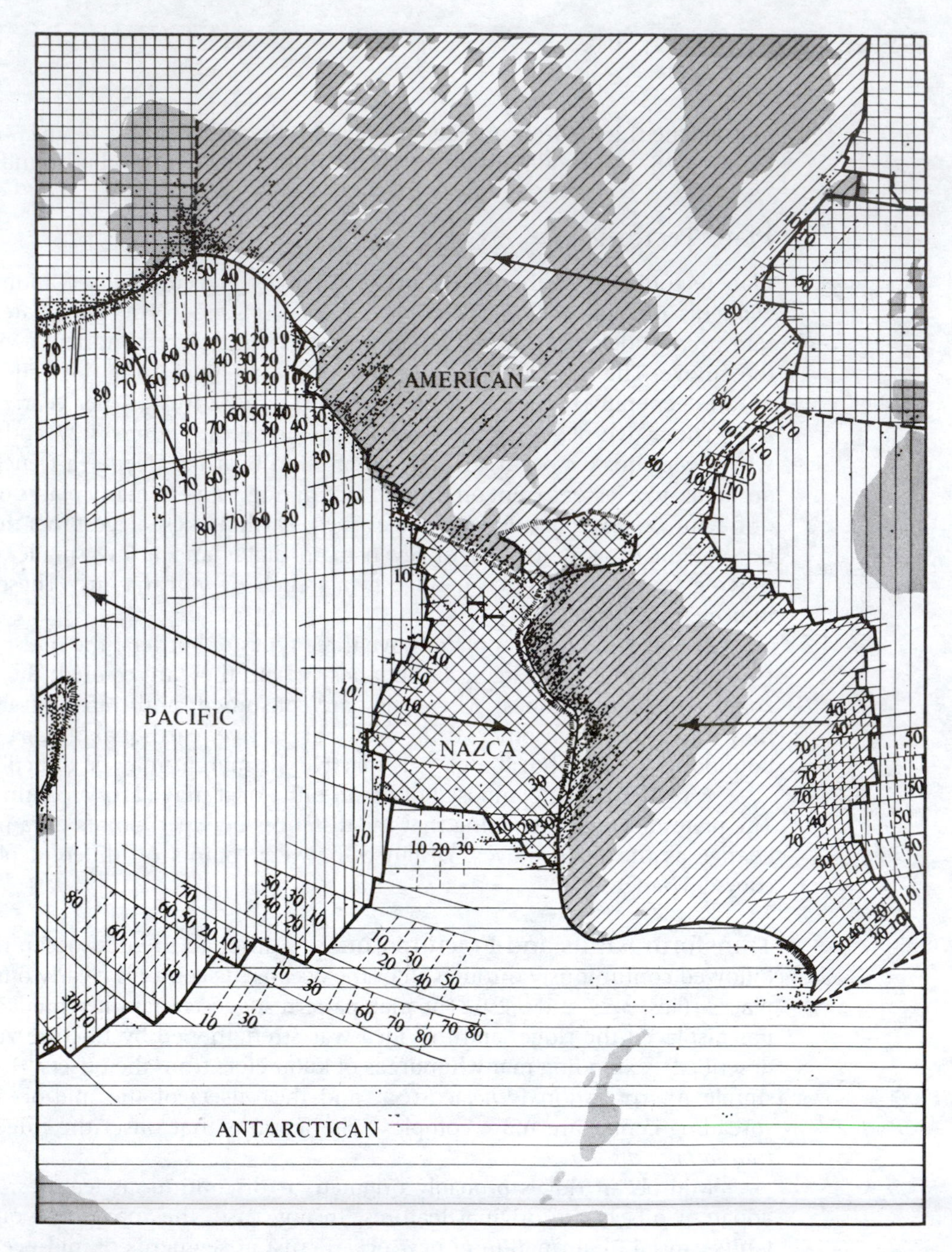

FIGURE 3.29
The plates. In general, each plate begins at a ridge (heavy line) and moves toward a trench (hachuring). Earthquake epicenters (dots) tend to outline the plates. Transform faults are shown by thin lines. The dashed lines show the ages (in millions of years), based on magnetic data. (The information and interpretation are from many sources, mainly from workers at Lamont-Doherty Geological Observatory.)

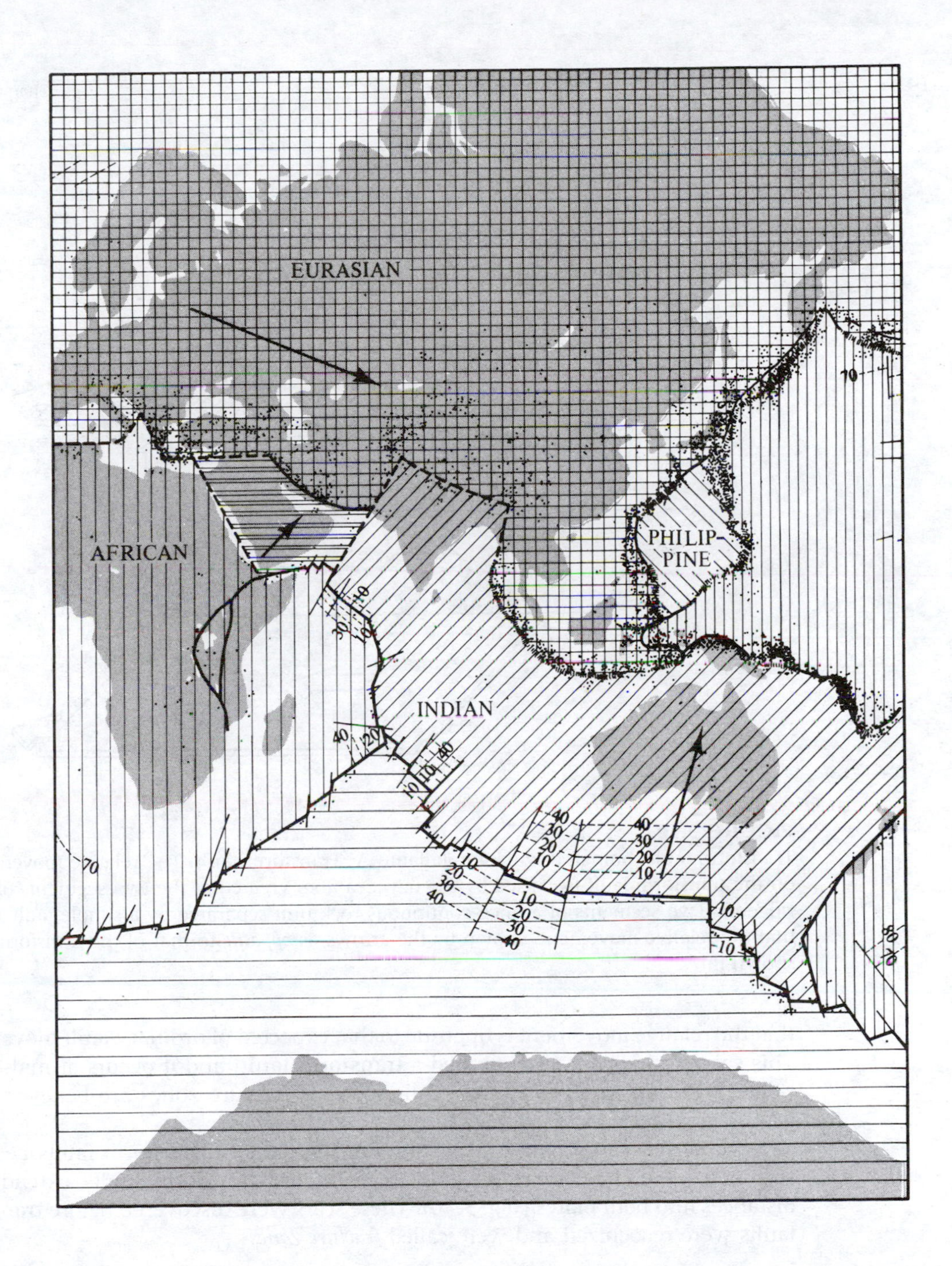
EURASIAN
AFRICAN
INDIAN
PHILIP-
PINE

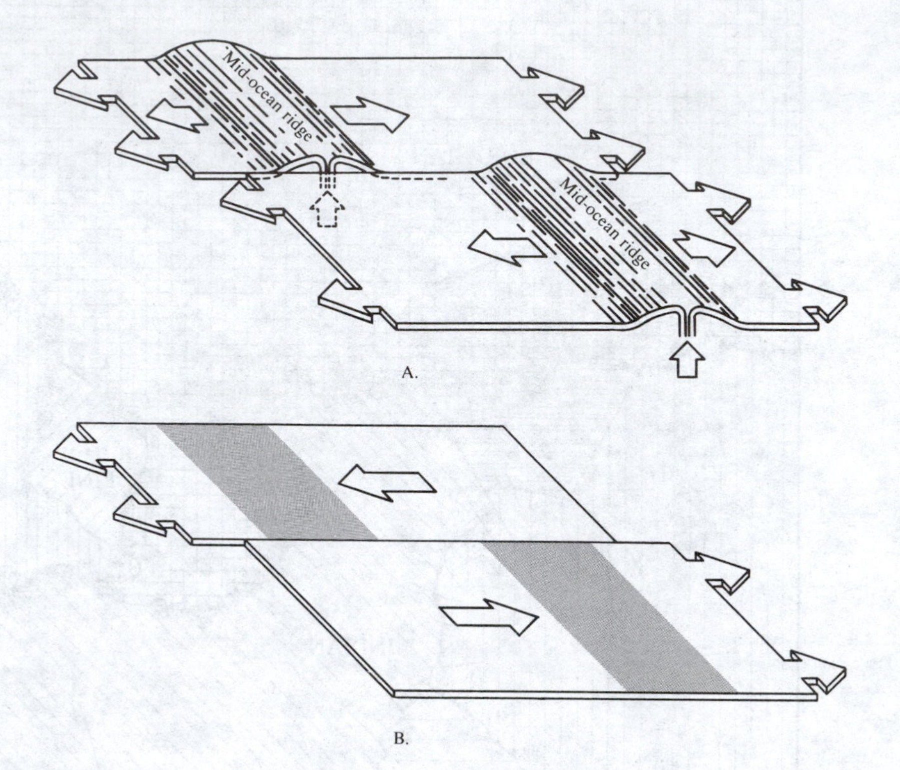

FIGURE 3.30
Transform fault compared with simple fault. A. Transform fault. The relative movement on the transform fault is caused by the new sea floor created at the two segments of the ridge. B. Two segments of a once-continuous rock unit separated by a simple fault. Note that the relative movement, shown by the arrows, is *opposite* to that of the transform fault shown in A.

that the relative movement is opposite to that expected of ordinary fault movement. This special kind of fault is termed a **transform fault,** and it occurs at mid-ocean ridges. The San Andreas Fault and the oceanic fracture zones are believed to be transform faults.

Note that earthquakes occur only on the parts of transform faults between segments of mid-ocean ridges. The scars of the transform faults extend long distances into both plates (Fig. 3.30). These scars were discovered before transform faults were recognized and were called *fracture zones*.

Plate Motions and Hot Spots The direction of movement can be determined from a map of the plate. Plate motion in general is from a mid-ocean ridge or spreading center toward a subduction zone. Transform faults reveal the actual direction because motion must be parallel to a transform fault. If the motion is not parallel to a transform, the plate would have to be either created or consumed at the transform. Most transform faults are at right angles to the spreading center, but plate motion can be at any angle from a spreading center, and a plate can move at any angle into a subduction zone.

Rate of movement of one plate relative to another is determined from magnetic stripes. The times of magnetic field reversal are known, so the width of the stripes divided by the time over which they formed gives the rate of movement of the plate at that point. Absolute rates of movement can be determined from the ages of islands and seamounts that are believed to form over fixed hot spots or mantle plumes.

Many volcanic areas are not near mid-ocean ridges or subduction zones. This suggests that hot material rises through the mantle at such places. These areas are called **hot spots,** and the hypothesized column of rising hot material is termed a **mantle plume.** The underlying idea is that at mid-ocean ridges, magma rises as sheets, and that the mantle plumes are similar columns. Some hot spots occur on continents (Yellowstone National Park is an example), and some, such as the Hawaiian Islands, form in oceans.

The mantle plumes associated with some hot spots appear to be very long-lived and to be fixed in location. A plate moving over such a place would have a line of volcanoes. The volcanoes should be progressively younger as the hot spot is approached, with active volcanoes at the hot spot.

The active volcanoes in the Hawaiian Islands are at the southeastern end of an island group. From there, going westward toward Midway Island, the basalt on the island volcanoes is progressively older (Fig. 3.31). Beyond Midway Island is the line of the north-trending Emperor Seamounts, and they continue the trend of older rocks to the north. These volcanic islands and seamounts are interpreted as having formed as the Pacific plate moved over the hot spot where the Hawaiian Islands are today. In this interpretation, the change in direction of the volcanoes would be caused by a change in direction of the movement of the Pacific plate.

The Cause of Plate Tectonics To understand plate tectonics, we need to know both the mechanism or process that creates and moves plates, and the energy source that drives the process. Convection currents within the mantle have long been suggested as the driving force of most internal processes, but as we shall see, other mechanisms also may be at work. The energy source also is in doubt, with radioactivity and deep, internal heat sources being the leading contenders.

Convection Convection currents have been proposed as the cause of most large-scale geological features. **Convection cells** *develop when a fluid is heated at the bottom. As the warmed fluid expands, it becomes less dense and tends to rise; denser, cooler fluid at the top moves down to take the place of the rising, warm fluid, as shown in Figure 3.32. In this process heat is transported upward by the movement of the warmed fluid.*

Although there is disagreement on the viscosity of the mantle, convection has been explored theoretically by a number of investigators who have established that it could exist in the mantle, even though the mantle is viscous enough to behave

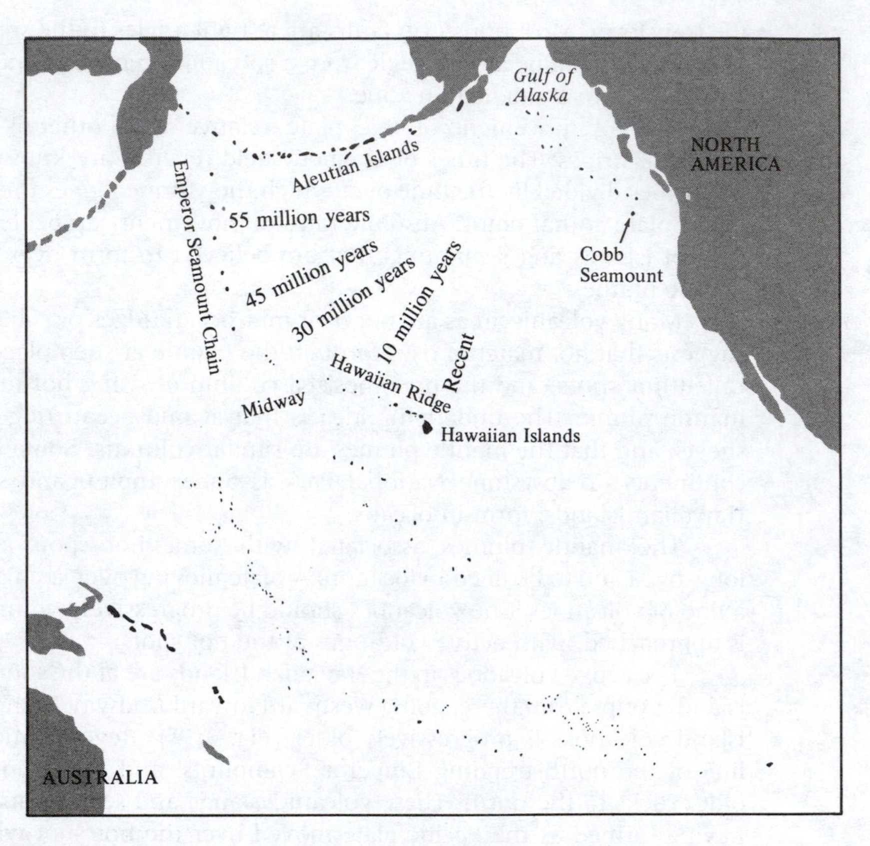

FIGURE 3.31
The line of seamounts and islands may have been formed as the Pacific plate passed over a hot spot. Approximate ages of the basalt are indicated.

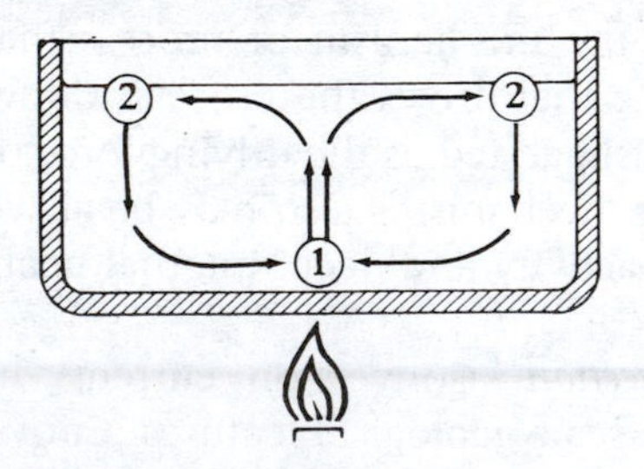

FIGURE 3.32
Two convection cells produced by heating at the bottom. Warm fluid at 1 rises and cool fluid from 2 falls to take the place of the rising warm fluid.

as a solid in transmission of earthquake waves. A substance may respond elastically to a rapid stress but may be deformed viscously by a stress applied over a long period of time. The latter case could occur in the mantle, especially in view of the temperatures and pressures there. Thus, it is possible for a solid to flow at temperatures well below the melting point. A possible example of this behavior is the slow movement (flow) of glacial ice under the influence of gravity, contrasted

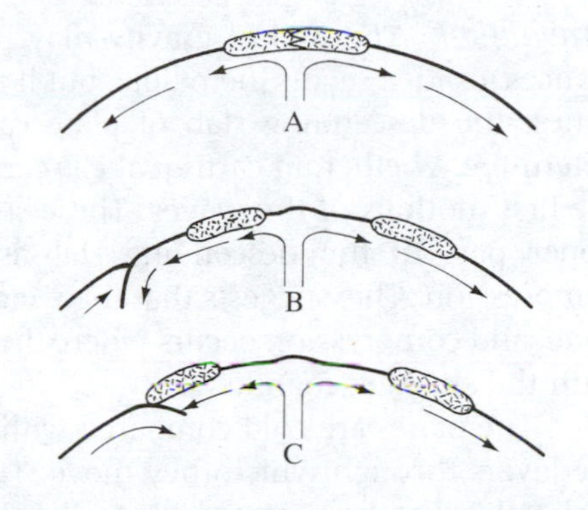

FIGURE 3.33
Possible role of convection currents in continental drift. Formation of ridge beneath continent could result in splitting of the continent and movement to a trench.

with its very different behavior when subjected to a rapid stress (shattering). Convection currents are believed to move at the rate of one to a few centimeters per year.

Many features of the mid-ocean ridges can be explained by convection currents. The high heat flow suggests upwelling convection currents in the mantle. Convection currents could cause the rise. The upwelling mantle material could melt partially as a result of the reduced pressure near the crust-mantle boundary, forming basalt. The basalt would become new ocean floor. Horizontal currents could cause sea-floor spreading, and descending currents could cause trenches and volcanic island arcs (Figs. 3.33 and 3.34).

FIGURE 3.34
View northeast across the Sinai Peninsula taken from *Gemini XI.* The southeast end of the Mediterranean Sea is at the upper left, and the north end of the Red Sea is at the lower right. The Sinai Peninsula is bordered by the Gulf of Suez and the Gulf of ‘Aqaba. The Gulf of ‘Aqaba is part of a linear depression on which the Dead Sea can be seen. The Red Sea may be caused by a new mid-ocean ridge that originated when Africa and Arabia separated, and the two gulfs shown here may be related to that fracture. (Photo from NASA)

Gravitation The pull of gravity may be the force that moves the plates. At first glance this idea seems incredible, but it is supported by seismic data. At subduction zones, the descending slab of plate causes many earthquakes. It is possible to determine whether an earthquake is caused by tension or compression by studying the first motions of the waves. These studies reveal that, at subduction zones, the upper part of the descending slab is in tension, and the deeper part is in compression. This suggests that the weight of the plate pulls it down into the weak zone, and compression occurs where the deeper parts of the plate come into contact with the stronger mantle.

The plates are cold compared with the upper mantle, so they are denser than the layers through which they move. This difference in density may be enough to make this mechanism work. At depth, the descending plate is heated by the surrounding hotter rocks, and the slab is assimilated into the mantle.

The height of the mid-ocean ridges also may provide a small gravity force. In this instance, the plate is simply sliding downhill. The ridge itself may be caused by expansion due to heating by the hot mantle material that rises there.

Gravitational pull is similar to convection in that both processes depend on differences in density. Indeed, they may both work together to cause plate tectonics. Gravitational pull may cause the plates to move, and convection in the mantle may help to propel the plates. If we knew the energy source, we might have help in deciding the relative importance of these possible mechanisms.

Energy Sources The energy that runs the earth's internal engine probably comes from two sources: unstable radioactive elements that spontaneously decay, releasing energy, and internal energy stored within the earth from the time of its formation. This energy is in the form of heat, and it is possible to measure the amount of heat coming from inside the earth.

Radioactivity is the source of at least some of the heat flow. Radioactive elements are concentrated in the crust and are not abundant in the mantle. The known radioactive content of crustal rocks of the continents is enough to account for the continental heat flow. The average heat flow from the oceans is about the same as the average from the continents. However, the thin layer of oceanic basalt does not contain enough radioactive elements to account for the oceanic heat flow. Therefore, some of the heat flow from the oceans must come from the mantle, and this is especially true of mid-ocean ridges where the heat flow is very high.

Radioactive elements are fairly evenly distributed in the continental crust; geologically active areas, such as volcanoes and mountain belts, are no more radioactive than other areas. Radioactive energy is enough to cause all geologic processes, but the distribution of radioactivity suggests that it is not the energy source, so other sources must be considered.

Some other aspects of radioactivity also should be considered. As radioactive elements decay, the total number of radioactive atoms is reduced; thus, the amount of radioactive heating has decreased throughout geologic time. This implies that if radioactivity causes plate tectonics, plate movements should be slowing. Another point is that if convection in the mantle is involved in plate tectonics, a deep heat source is necessary because convection requires heating from below. Radioactivity is concentrated in the crust. Finally, analysis of heat-flow data strongly suggests that well over half of the heat flow, even on the continents, comes from below the crust, and, of course, most of the heat flow in the oceans also comes from deep sources.

Some of the heat reaching the surface must come from deep in the earth. The deep source of heat is believed to be the liquid core. The heat could be released by crystallization of the core, and the inner core, believed to be solid, may be slowly enlarging. The liquid core also is believed to be the source of our magnetic field. Internal heat also could come from recrystallization in the mantle, but we know very little about either the composition or the physical conditions there.

Differences Between Oceans and Continents

The relief features of the earth are summarized in Figure 3.35. This diagram shows that the very high and very low portions of the earth are very small in area, and two levels, the continents and the ocean basin floors, make up most of the surface. The

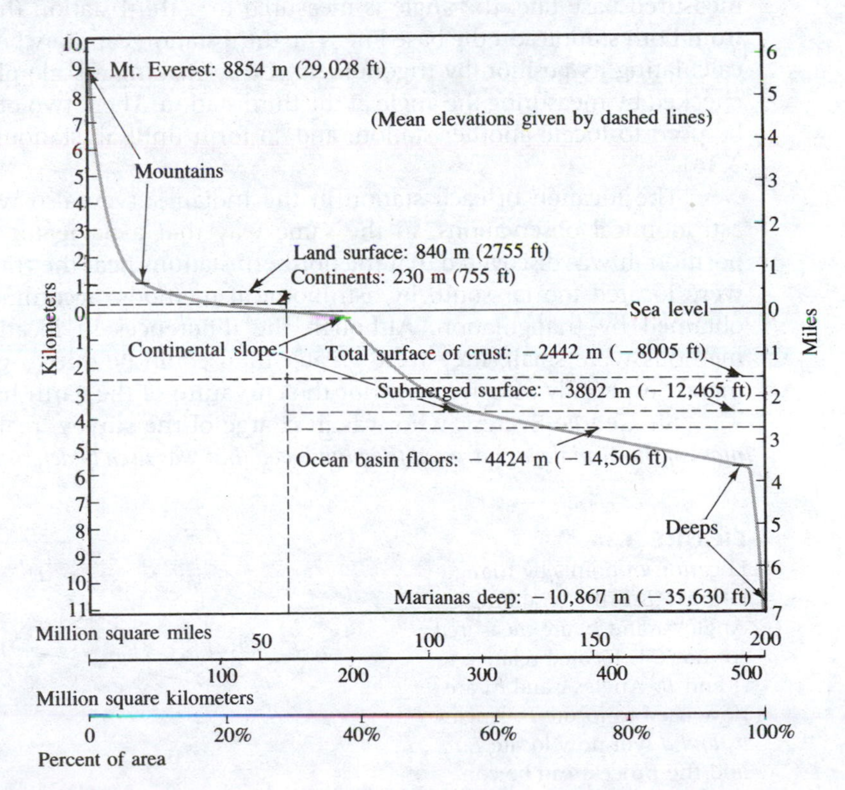

FIGURE 3.35
Elevations of the earth's crust.

reason for the two distinct levels was one of the major problems of geology. Note that the difference between the highest and the lowest points is about 19.3 km (12 miles), and this is very small compared with the radius of the earth (19.3 km ÷ 6364 km × 100 = 0.3%), so that the earth is smoother than a billiard ball. This comparison shows that the features we observe on the surface are very slight compared with the earth as a whole.

Certainly the differences between ocean and continent are the most pronounced contrast on the earth's surface. One important contrast is in rock type—the ocean basins are composed mainly of basalt; the continents of igneous and metamorphic rocks of granitic composition. To understand how these compositional differences might account for the two main levels in the elevation diagram (Fig. 3.35), we must turn to a different type of study.

Surprisingly, an important clue came from a surveying error made about the middle of the nineteenth century when the English were mapping India. Their methods of surveying were much the same as our modern way; only the instruments were different. Because it is easier and more accurate to measure angles than distances, most surveys use **triangulation.** To use this method, a base line is accurately measured, generally by actually taping it. From each end of this measured base line, the angle is measured to a third station that must be visible from both stations on the base line. The third station can now be located, either by calculating its position by trigonometry or by making a scale plot. The triangle is checked by measuring the angle at the third station. Then, two of these stations can be used to locate another station, and so forth until all stations are located (Fig. 3.36).

The location of each station in the Indian survey also was determined by astronomical observations, in the same way that a navigator at sea locates his position. It was discovered that the northern stations near the Himalaya Mountains were located too far south by astronomical methods, according to the locations obtained by triangulation. Although the differences in location by these two methods were small, they were greater than could be due to errors in measurements, especially as corrections for the curvature of the earth had been applied.

Sir George Everest, who was in charge of the survey, realized that *the great mass of the Himalayas attracted the plumb bob that was used to determine the zenith in the*

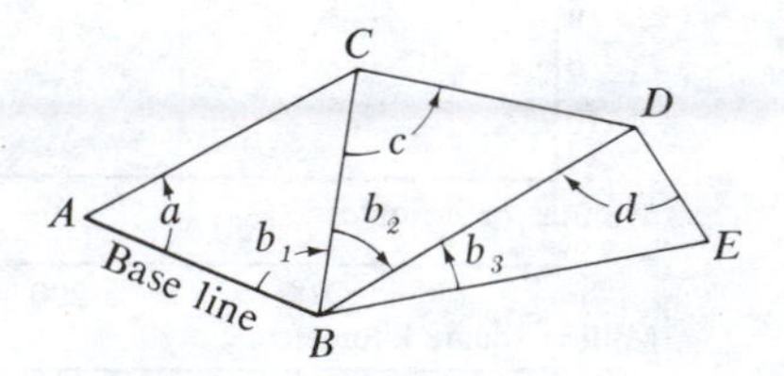

FIGURE 3.36
Location of points by triangulation. *AB* is the base line. Angles *a* and b_1 are measured so that *C* is located relative to *A* and *B*. Angles *c* and b_2 are now used to locate *D*. Angles b_3 *and d* will now locate *E*, and the process can be continued.

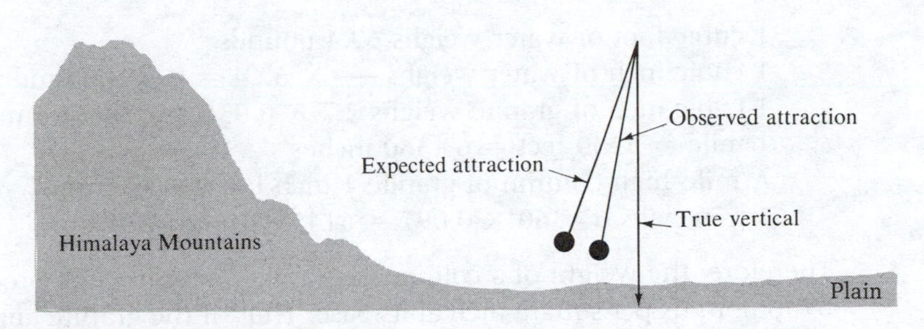

FIGURE 3.37
The gravitational attraction between the Himalaya Mountains and the plumb bob causes the plumb bob to be deflected from the true vertical. The deflection is less than expected, suggesting that mountains are underlain by light rocks.

astronomical determination of location. A level bubble and a plumb bob are basically the same, and one or the other must be used to establish the zenith from which astronomical measurements are made. Knowing the size of the mountain range and using Newton's law of gravitation, he calculated the expected amount of attraction of the plumb bob by the range and discovered that the observed deflection was much less than the expected deflection (Fig. 3.37). This discrepancy implies either that the mountains are composed of rock lighter than the plain, or, if *both mountains and plains are composed of the same rock type, that the rock under the mountains is lighter. The latter case,* **roots of mountains,** *is supported by seismic data and geologic mapping* (Fig. 3.38).

This theory suggests that the granitic continents, like the mountains, may stand higher because they are thicker and less dense than the basaltic oceans, granite being less dense than basalt. To find a possible layer in which the crustal rocks may float, we will begin by considering the strength of the crustal rocks.

Laboratory tests show that granite, the main rock composing the continents, has a compressive strength of about 22,000 pounds per square inch. The specific gravity of granite is about 2.7; that is, it weighs 2.7 times as much as an equal volume of water. From these data we can calculate how high a column of granite can support itself; that is, at what height the granite will weigh more than its compressive strength.

FIGURE 3.38
The root of a mountain range.

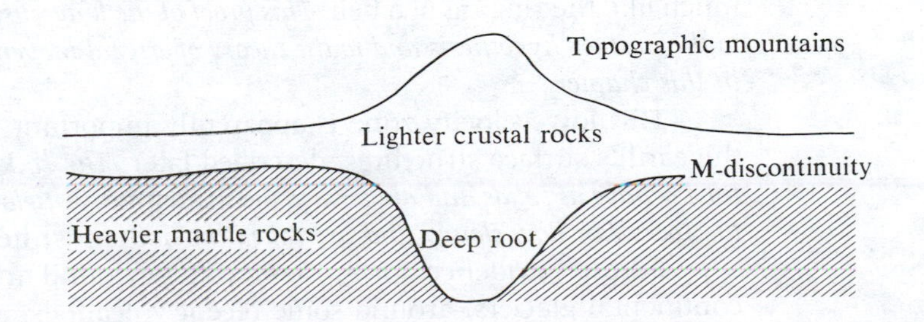

1 cubic foot of water weighs 62.4 pounds
1 cubic inch of water weighs $\frac{1}{1728} \times 62.4 = 0.036$ pound
1 cubic inch of granite weighs $2.7 \times 0.036 = 0.097$ pound
1 mile = 5280 feet = 63,360 inches
A mile-high column of granite 1 inch in cross section
weighs $63{,}360 \times 0.097 = 6145.92$ pounds

Therefore, the weight of a column of granite one mile high produces a pressure of 6146 pounds per square inch at its base. Thus, if the granite approaches four miles in thickness, its compressive strength is exceeded, and the base will crumble under its own weight. The continental crust exceeds this thickness, and some mountain ranges are higher. How is this possible? A factor we have so far neglected, but which will not basically alter the problem, is the change in strength of granite due to the increase in temperature and pressure at depth. The specific gravity of rock will increase due to pressure, the pressure will tend to strengthen the rock, and the higher temperature will weaken it. The total of these effects is to increase greatly the height of the column of rock that is self-supporting, but clearly a limit will be reached. Because the conditions deep in the crust and below it cannot be duplicated in the laboratory, and because the exact composition of the rocks is not known, an accurate height cannot be calculated. It is clear, however, that the continents cannot be supported by their rigidity.

Apparently, then, the high temperature and pressure at some depth make the rocks plastic, and the crust floats on these plastic rocks. Because the rocks in this zone are weak, probably because they are near the melting point for the pressure there, this could be the zone where magmas are generated. *The theory that the crust floats on this viscous substratum is called* **isostasy.** The depth at which the rocks become plastic is not easy to determine. *It is not at the Mohorovičić discontinuity,* which marks the base of the crust, but is between 80 and 200 km (50 and 125 miles) below the surface, and so is in the mantle. Apparently, below this depth the increased pressure increases the strength of the rocks, and they behave as solids, at least to rapid deformations such as seismic waves generated by deep earthquakes. *This plastic zone is the* **low-seismic-velocity zone** *(between about 80 and 200 km)* that was, on the basis of sparse seismic data, thought to exist. Study of the seismic waves generated by underground nuclear blasts revealed shadow zones due to refraction of waves by the low-velocity zone (Fig. 3.39). This layer also was shown by the intense Chilean earthquake of 1960, which set the zone above this layer vibrating, much like the ringing of a bell. *This proof of the long-suspected weak layer adds credence to the isostasy hypothesis and to the theory of crustal movement and deformation described in this chapter.*

This low-velocity zone is apparently important in other processes that form the earth's surface structures, described later. *The rocks above the low-seismic-velocity zone behave as solids and are termed the* **lithosphere.** *Below the low-seismic-velocity zone, the rocks can flow slowly and are called the* **asthenosphere** (Fig. 3.40).

Further evidence for isostasy can be found in areas recently covered by continental glaciers, around some recent volcanoes, and by measurements of the

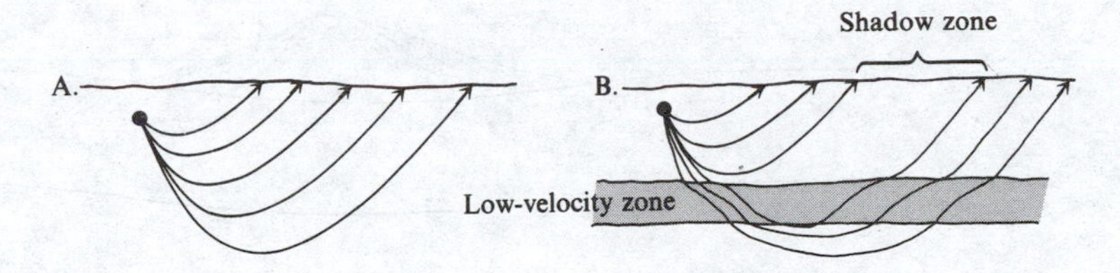

FIGURE 3.39
The recognition of the low-velocity zone in the upper mantle. A. Shows the paths of seismic waves with no low-velocity zone. B. Shows diagrammatically how refraction in the low-velocity zone produces a shadow zone at the surface. (In both diagrams, for simplicity, the refraction that occurs at the M-discontinuity is omitted.)

gravitational attraction near most mountain ranges. Parts of the earth that were covered by thick accumulations of ice during the recent glacial period apparently were depressed by the weight, and now that the ice has melted, they are rising. This rising may be due in part to elastic or plastic compression of the crust, but calculation suggests that these effects are not enough to account for the postglacial uplifts and that isostasy is involved. The Hawaiian Islands are a group of young volcanoes, and their weight has apparently depressed the crust, creating a moat of deeper water surrounding them, as shown in Figure 3.41. On the continent, erosional debris would mask a similar occurrence where a rapidly emplaced

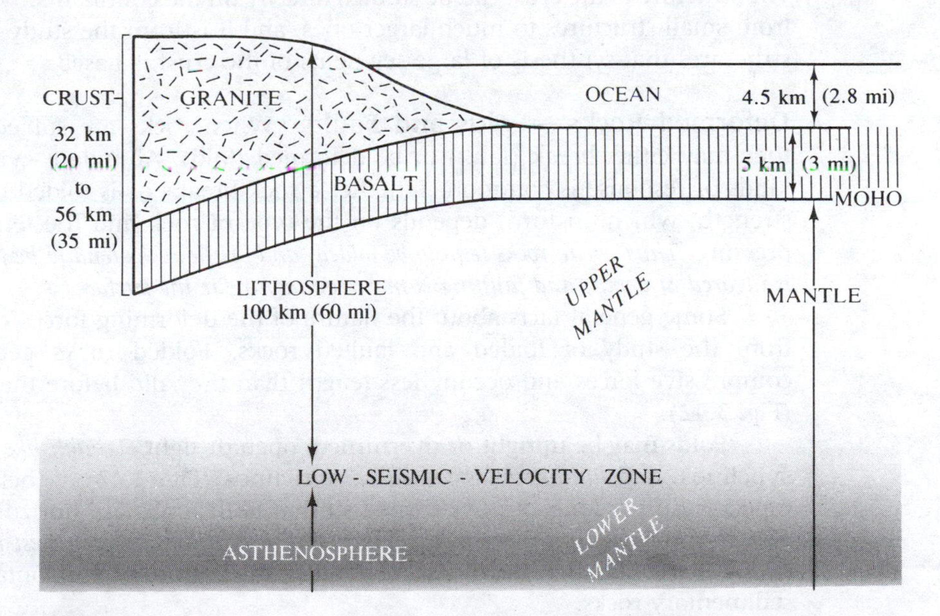

FIGURE 3.40
The crust and upper mantle form the lithosphere, in which the rocks behave as solids. In the asthenosphere, below the low-seismic-velocity zone, the rocks flow.

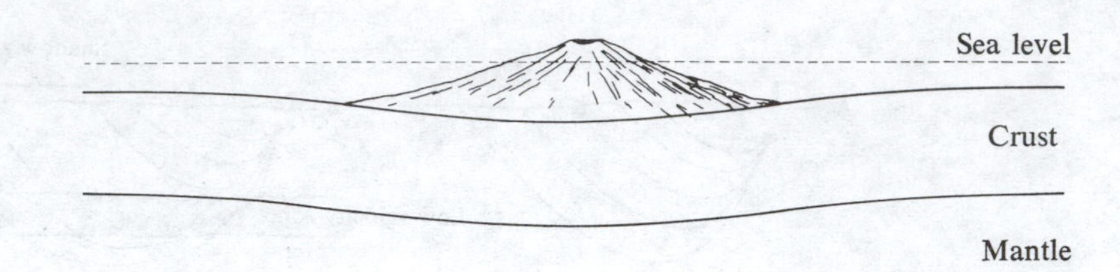

FIGURE 3.41
Depression of the crust into the viscous mantle by addition of a weight such as a volcano. As erosion removes the weight, the crust would return to its original position. The Hawaiian Islands and some other volcanic piles in the Pacific Ocean have apparently depressed the crust in this manner, because they are surrounded by a ring of deeper water.

volcano depresses the crust. Measurements of the gravity force—which of course is affected by the mass distribution under the point—when corrected for such effects as elevation and nearby mountain mass, show that most of the topographic features of the earth are compensated at depth, and thus tend to prove the hypothesis of isostasy. This suggests that movements in the plastic zone that tend to restore isostatic balance are fairly rapid.

Structure of the Continents

The structure of the crust can be studied directly on the continents. We will proceed from small structures to much larger ones, and it is from the study of these lesser structures that synthesis of large segments of the crust is based.

Deformed Rocks—Folds and Faults When rocks are subjected to forces, they may either break (fault) or be deformed (fold). Abundant evidence of both kinds of response is common. Whether a rock breaks or is folded depends on its strength, which, in turn, depends on the type of rock and the temperature and pressure. *Thus, weak rocks tend to be folded, and brittle rocks tend to be faulted. Folding is favored at depth, and faulting is more common near the surface.*

Some general facts about the nature of the deforming forces can be learned from the study of folded and faulted rocks. Folded rocks generally record compressive forces and occupy less length than they did before they were folded (Fig. 3.42).

Folds may be upright or overturned, open or tight. *Trough-like folds are called* **synclines,** and *hill-like folds are called* **anticlines** (Fig. 3.43). A better definition, especially if the folds are overturned so that both limbs are horizontal, is that *in anticlines the oldest rocks are in the center and in synclines the youngest rocks are in the center.* Remember that the oldest rocks are at the bottom of horizontal undeformed sedimentary rocks.

Faulting may record either shortening or lengthening of the rocks, as shown in Figures 3.44, 3.45, and 3.46. Faults also may have horizontal movement or a

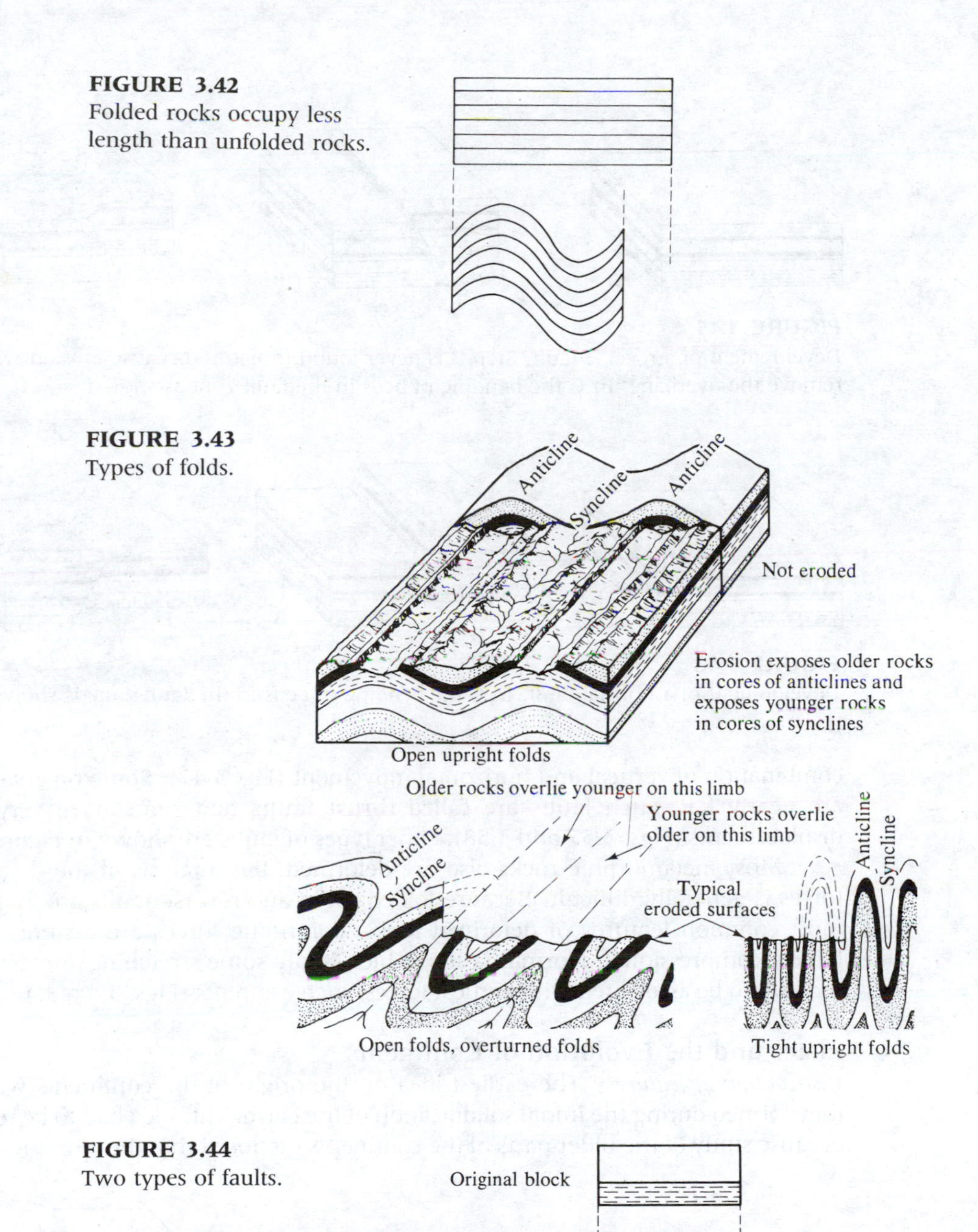

FIGURE 3.42
Folded rocks occupy less length than unfolded rocks.

FIGURE 3.43
Types of folds.

FIGURE 3.44
Two types of faults.

Original block

Compressive or reverse fault shortens block

Tensional or normal fault lengthens block

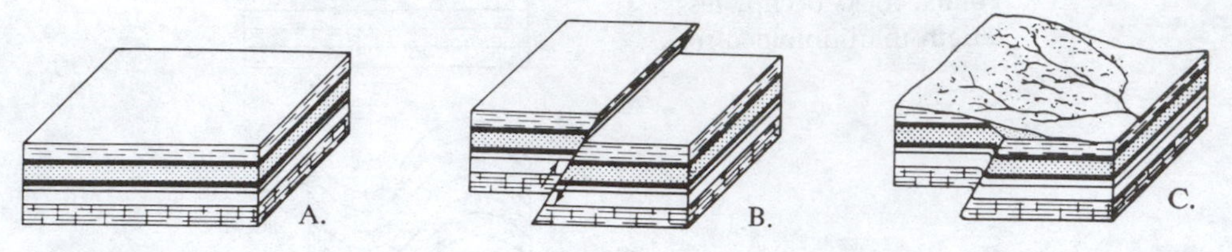

FIGURE 3.45
Development of a reverse fault. Step B is never found in nature because erosion would remove the overhang. In C the bending of beds in the fault zone is shown.

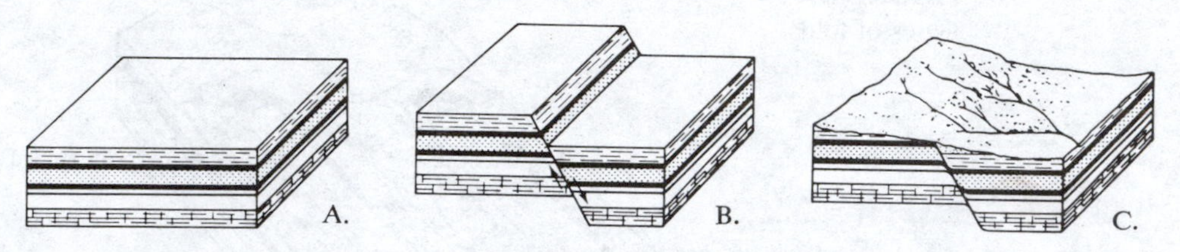

FIGURE 3.46
Development of a normal fault. In C the bending of beds in the fault zone is shown.

combination of vertical and horizontal movement (Fig. 3.47). Some **reverse faults** are nearly flat. Such faults are called **thrust faults** and some have very large displacements (Figs. 3.57 and 3.58). Other types of faults are shown in Figure 3.30.

Most metamorphic rocks also are deformed, but analysis of the deforming forces is generally difficult. Because folded rocks and reverse faults are by far the most common features of deformed rocks, deforming forces are assumed to be largely compressional. **Normal faults,** which imply some stretching, are generally thought to be associated with vertical forces, such as uplifts (Figs. 3.44, 3.46, 3.47).

Plates and the Evolution of Continents

Distribution of features The earliest idea on the origin of the continents was that they formed during the initial solidification of the earth. This idea had to be rejected because study of the older parts of the continents disclosed that the oldest rocks are

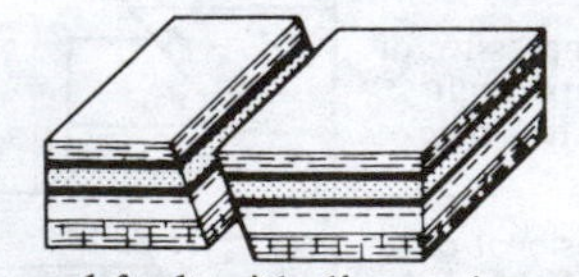

A. Normal fault with diagonal movement

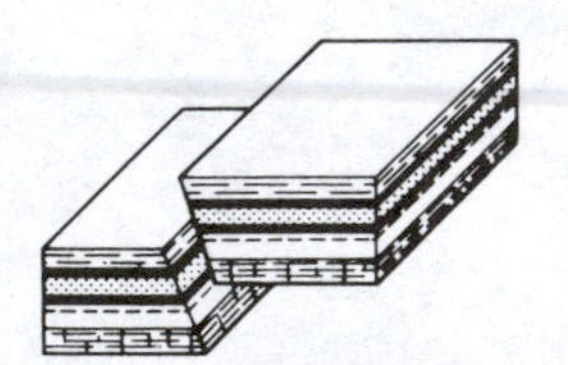

B. Reverse fault with diagonal movement

FIGURE 3.47
Faults with oblique movements.

metamorphosed sedimentary rocks that were themselves the products of erosion of a still earlier terrain. The original crust of the earth has yet to be found, even with the radioactive age determinations of today; however, these radioactive dates have revealed that, in a general way, the oldest rocks are surrounded by belts of younger rocks, and these by still younger rocks. Because each of these belts of younger rocks has structures similar to parts of some of our present mountain ranges, the belts may be the eroded stumps of older mountain ranges (Fig. 3.48).

This discovery suggests that the present surface processes have been going on since the formation of the crust, and that the continents may have formed by accretion. The oldest rocks yet discovered are about 4 billion (4000 million) years old, which suggests a total age of the earth of between 4.5 and 5 billion years.

The continents are composed of igneous and metamorphic rocks of granitic composition, overlain at most places by a thin veneer of sedimentary rocks. In general, in the interior of a continent, the surface sedimentary rocks are thin and relatively undeformed; but as the oceans are approached, the layers are thicker, volcanic and intrusive rocks become more abundant, and folding and faulting

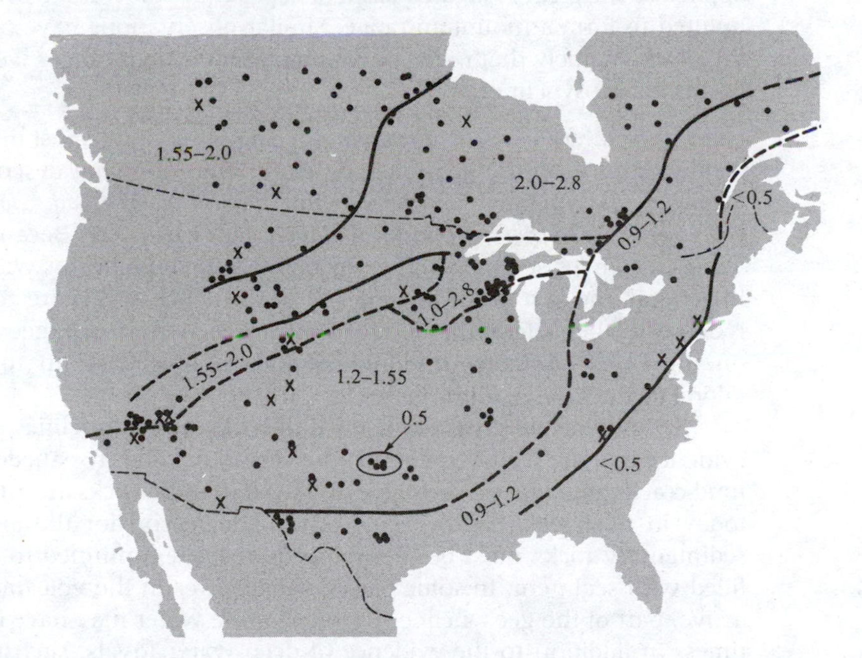

FIGURE 3.48
Distribution of ages in crystalline rocks from the central part of North America. Circles: ages that are within the limits specified for a given zone on the map; crosses: ages that are outside the limits. Age limits are given in billions of years. (After G. R. Tilton and S. R. Hart, "Geochronology," *Science,* Vol. 140, 26 April 1963, p. 364, © 1963, American Association for the Advancement of Science)

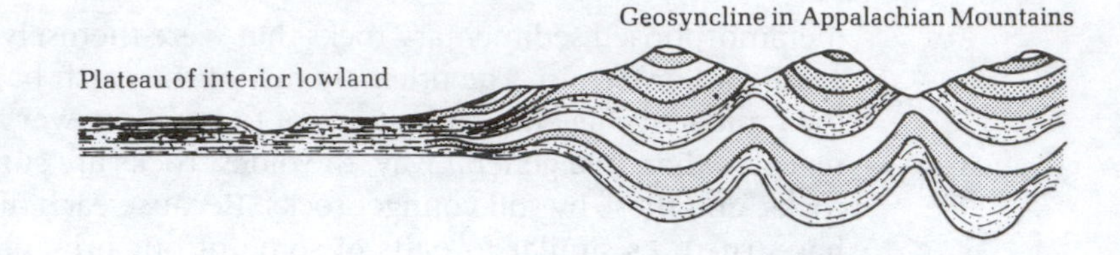

FIGURE 3.49 Sedimentary rocks of the same age are thicker and more clastic in mountain ranges than in the interior of continents.

become conspicuous. The ages shown in Figure 3.48 are those of the underlying granitic rocks.

Modern ideas on the formation of mountains began more than a hundred years ago. In 1859 James Hall, an American geologist who made important studies of fossils, discovered that rocks in the northern part of the Appalachian Mountains were much thicker and more clastic than were rocks of the same age located in the interior lowlands. In other words, the present mountains are on the site of a much deeper basin than is the area to the west (Fig. 3.49). This observation was extended by James Dwight Dana, who recognized that it was true of the whole Appalachian Mountains. His studies culminated in 1873 when the name **geosyncline** was applied to the great elongated basin of deposition that was later folded, faulted, and uplifted to form a mountain range. Similar observations have been made all over the world, namely that many of the present mountain ranges were once elongated basins, or geosynclines.

Geosyncline theory—origin of mountains The observations just discussed have been combined to form the geosyncline theory. The following description is of a very idealized geosyncline. *The* **geosyncline** *begins as a broad, elongated downwarp, generally at the edge of a continent, that is flooded by seawater.* Because the geosyncline acts as a trap for sediments, the **sedimentation stage** begins with this downwarp. In general, the part of the geosyncline near the ocean is more active than the rest, is depressed deeper, and is the site of volcanic activity. These geosynclines are large, on the order of many hundred kilometers long and up to several hundred kilometers across.

In spite of the great thicknesses of rocks in geosynclines, there is abundant evidence that the water was at no time very deep. The cross-bedding, ripple marks, mud cracks, and types of fossil life displayed in these rocks are all features that form today in relatively shallow water. Thus, the reason for the great thicknesses of sedimentary rocks must be that the geosynclines continued to sink as they were filled with sediment. In some places, the features in the volcanic, or at least more active, part of the geosyncline suggest that the water may have been deep there at times. In addition to the evidence of deep-water fossils, such features as graded bedding and channeling due to mud and muddy water sliding down slopes suggest some deep water. Such slides are known to occur today on the slope between the continental shelf and the deep ocean.

Differences in environment are shown by differences in sedimentation. As shown in Figure 3.50, part 1, the geosyncline gradually tapers to form the shelf that

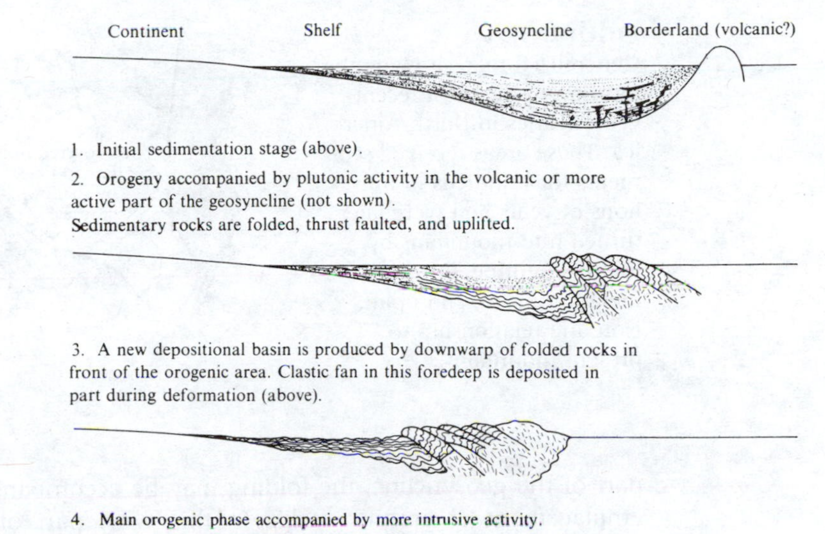

FIGURE 3.50 Stages in development of a geosyncline. Compare with Figure 3.53.

borders the continent. In most cases the continent has little relief; this is shown by the sediments (shale, limestone, and fairly pure quartz sandstone) that form on the shelf. Such sediments reveal rather complete weathering. This same suite of sedimentary rocks also develops nearby in the geosyncline, but here the rocks are much thicker because of the greater downsinking. The volcanic part of the geosyncline, on the other hand, is the scene of more rapid sedimentation as a result of deeper and more rapid downwarping. In addition to volcanic rocks, the sediments here are rapidly deposited, poorly sorted, quickly buried clastic rocks, such as impure shales, dark sandstones with clay matrix (graywacke), and coarser clastic rocks.

The source of these "dumped-in" clastics is a problem. Some of the material may come from the continent, but the volume generally requires a source on the other side of the geosyncline. The problem is further complicated by the later igneous and metamorphic activity which obscures the early situation in this part of the geosyncline.

The sedimentation phase of a geosyncline ends when it stops sinking and receiving sediments, generally after 9000 to 15,000 m (30,000 to 50,000 feet) of sedimentary rocks have been deposited. The next phase is the **orogenic phase,** *during which (1) the sediments are deformed to make mountain structure and (2) the volcanic part of the geosyncline may be changed by igneous and metamorphic activity into a rigid mass of granitic composition, thus perhaps becoming a part of the continent* (Fig. 3.50).

The orogenic phase of the geosyncline is probably the most interesting, most important, most studied, and least understood phenomenon in geology. At this time the geosyncline is folded, apparently by compressive forces. In the volcanic

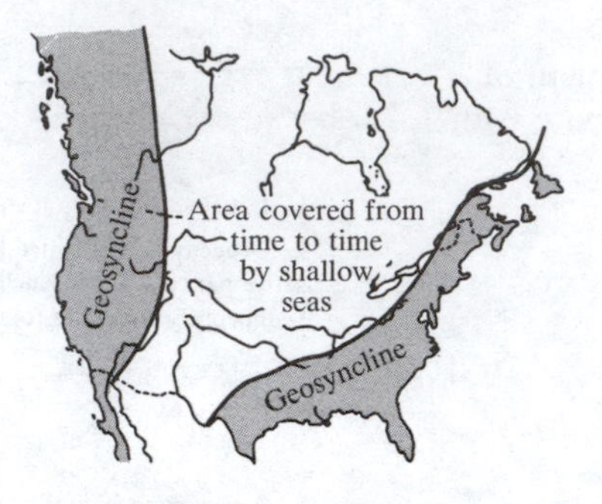

FIGURE 3.51
Generalized map showing the location of the most recent geosynclines in North America. These areas received sediments for hundreds of millions of years and were later turned into mountains by folding, faulting, intrusion, metamorphism, and uplift. Note the relationship to present mountains.

part of the geosyncline, the folding may be accompanied by metamorphism and emplacement of granitic rocks. In this way, part of the geosyncline may be transformed into a rigid mass. It should be noted that intrusive activity is not always confined to only the volcanic part of the geosyncline, and at places parts of the volcanic geosyncline may not be affected at all. At a few places batholiths, generally small ones, occur in nongeosynclinal areas. After the early orogeny, a new depositional basin generally develops over the border between the old geosyncline and the shelf or continent, and a new sedimentation cycle may ensue. This cycle may be followed by other episodes of folding, faulting, and thrust faulting in this or other parts of the geosyncline. Metamorphism and emplacement of batholiths may or may not accompany these orogenic events. These events may be spread over a few hundred million years. The thrust faults that may move thin plates of brittle rocks many kilometers are among the most difficult structures to explain, and several theories have been suggested. Some of these theories use sliding under gravity as the motivating force, because analysis shows that the rocks are not strong enough to transmit the compressive forces necessary to move the relatively thin plates without crumbling.

During the post-orogenic phases, it is probable that the geosynclinal areas are uplifted to form mountain ranges (Fig. 3.51). The geosynclinal deformation produces mountain structures, but topographic mountains do not exist until this uplift occurs. Perhaps the production of granitic rocks in the geosyncline leads to the uplift after whatever caused the initial downwarp has ceased to operate. Then the light granitic rocks would tend to float up under the influence of isostasy.

Once a mountain range exists, isostasy may control its future to a large extent. As soon as it rises, a mountain range is subjected to erosion. As erosion removes weight from the mountains, the buoyancy of the root probably causes the range to rise, as shown in Figure 3.52. This process should go on until the root has risen to the level of the rest of the continent and the topographic mountains have been reduced by erosion to the level of the rest of the continent. In this manner a new section of the continent may be formed. The rocks of this section would display

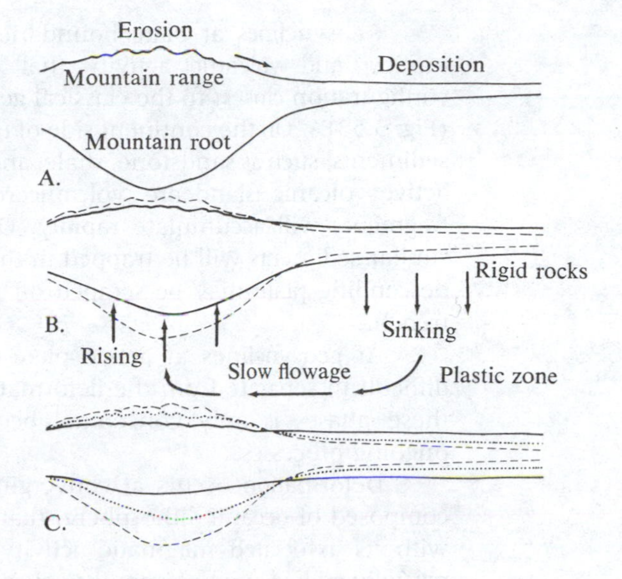

FIGURE 3.52
Removal of a mountain root by erosion and isostasy. Erosion removes material from the mountains, thereby lightening them. This can result in isostatic uplift that reduces the depth of the root. The place where the eroded material is deposited also may be depressed by the additional weight. Several stages in the process are shown in the diagram.

mountain structures, and this process could produce an age distribution similar to that in Figure 3.48.

Geosynclines and Plate Tectonics Uniformitarianism is the cornerstone of modern geology. If geosynclines are important in the geologic past, we should be able to recognize and study present-day geosynclines. We should be able to see where sediments are accumulating and where they are being deformed.

The simplest model of a geosyncline is the continental shelf–slope type of continent-ocean border (Fig. 3.53A). At typical continental shelf–slope areas, sandstone, limestone, and some shale are deposited on the continental shelf; these rocks, with more shale, form the thick sedimentary accumulations of the continental slope. All of these rocks are deposited in shallow water. This type of geosyncline can form only where the continent-ocean border occurs within a plate. Other geosynclines form at continent-ocean borders that are also margins of plates.

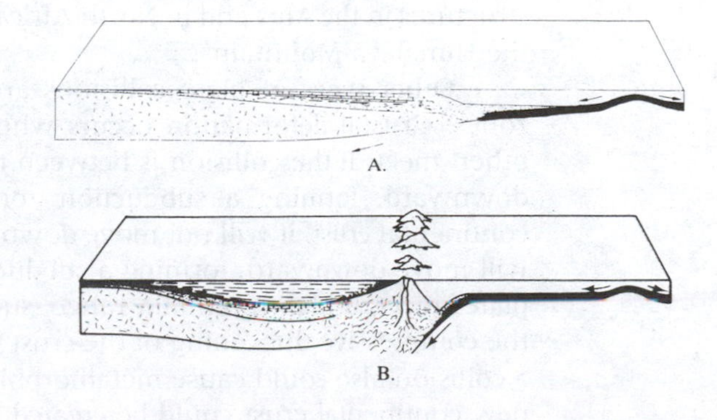

FIGURE 3.53
Two possible models of geosynclines associated with sea-floor spreading.
A. Continental shelf and slope.
B. Volcanic arc. B shows the origin of the igneous rocks from melting of a descending oceanic plate.

Geosynclines at plate boundaries are more complex because of the plate motion and volcanic activity that can result in a number of forms. The configuration closest to the classical geosyncline is the offshore volcanic island arc (Fig. 3.53B). On the continent side of this basin, typical continental shelf and slope sediments, such as sandstone, shale, and some limestone, will accumulate. Near the active volcanic island arc, volcanic rocks and sediments composed of volcanic fragments will accumulate rapidly. On the ocean side of the volcanic islands, similar sediments will be trapped in the trench. The ocean-floor sediments on the descending plate may be scraped off the plate and so also accumulate near the trench.

At geosynclines at active plate margins, the sedimentation phase is very difficult to separate from the deformation or orogenic phase. The separation into these phases is only conceptual because sedimentation and deformation are ongoing processes.

Deformation occurs at converging plate boundaries. If one of the plates is composed of oceanic lithosphere, that plate will descend and a subduction zone with its associated magmatic activity will be initiated. The descending slab is partially melted and becomes the source of magma. The rising magma and probably the active hot fluids derived from moist sediments on the melting plate cause metamorphism of the rocks above the subduction zone. In this way rocks of continental composition are formed. If the subduction zone exists for a long enough time, the volume of the rising magmatic material becomes large and sedimentary rocks farther inland are deformed by folding and faulting, especially thrust faulting. The rising magma may form an arch, causing thrust faulting by gravity gliding even farther inland (Fig. 3.54). The deformation phase ends when, for some reason, subduction ceases.

Collisions The deformation at a convergent plate boundary can be carried even further. If subduction occurs at the edge of a continent, and the plate being subducted is narrow and carries a continent on it, that continent will collide with the subduction zone.

Continent-continent collisions cause more intense compressional deformation than does subduction. The collision of Africa and Europe produced the thrust structures in the Alps and in North Africa. The collision of India with Asia produced the Himalaya Mountains.

Other types of plate collisions are possible, and they produce subduction zones. Crustal deformation occurs where two rigid plates, moving toward each other, meet. If the collision is between two plates of oceanic crust, one will move downward, forming a subduction zone (Fig. 3.55A). If one of the plates is continental crust, it will not move down because of its buoyancy; the oceanic plate will move downward, forming a subduction zone (Fig. 3.55B). If two continental plates meet, a high mountain range, such as the Himalayas, will form because of the compressive thickening of the crust (Fig. 3.55C). The heat and friction of such a collision also could cause metamorphism and generation of magma. In this way new continental crust could be created.

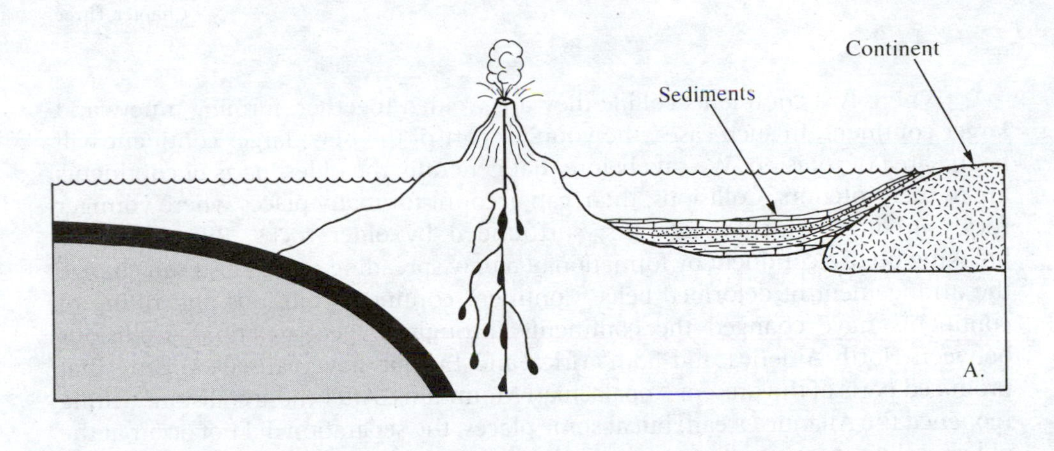

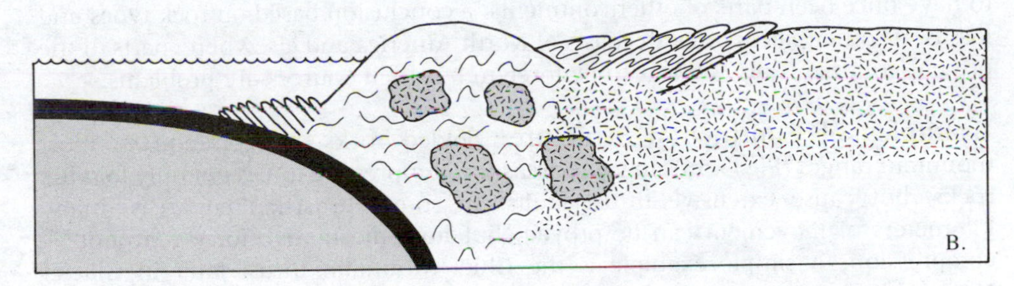

FIGURE 3.54
Subduction may cause deformation of geosynclinal rocks far inland from the volcanic arc.

FIGURE 3.55
Types of collisions of plates.
A. Ocean-ocean.
B. Continent-ocean.
C. Continent-continent.

Sea level

A. Ocean-ocean plate collision

Sea level

B. Continent-ocean plate collision. Ocean goes under continent.

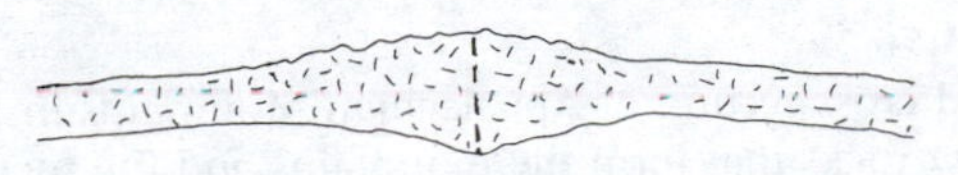

C. Continent-continent plate collision

Where two continents collide they are welded together, forming a new and larger continent. In such cases, the younger part of the new, larger continent will be the area of collision. We saw before that generally the oldest parts of continents are in their interiors. Collisions, then, can account for many places where younger rocks are in continental interiors, surrounded by older rocks. Rifting, or the breaking up of a continent by formation of a new spreading center, also can change the arrangement of deformed belts. Continent-continent collisions and rifting of continents have changed the continents throughout geologic time. Collisions between North America and both Africa and Europe have caused orogeny that produced parts of the present Appalachian Mountains. After these collisions, rifting reopened the Atlantic Ocean; but at some places, the separation did not occur at the old shoreline. As a result, some areas in New England and Newfoundland appear to have once been parts of other continents, a conclusion based on rock types and fossils. At other places on both coasts of North America and elsewhere, parts of the present continents also appear to be foreign, but their sources are problems.

Examples of Continental Structure Folded rocks are common in most mountain ranges (Fig. 3.56). In some areas the compressive forces not only fold the rocks, but cause extensive thrust faults. In some thrust-faulted areas, many kilometers of movement can be proved; but here the motive force may not be compression. A simple example is the Chief Mountain thrust fault in Glacier National Park, Montana, where the upper plate of strong, older sedimentary rocks has moved a minimum of 19 km (12 miles) (probably much farther) over weaker, younger sedimentary rocks (Fig. 3.57). Both north and south of here, the mountains have many such thrust faults. Similar structures have been mapped at many other places, such as northern Nevada and in the central part of the Appalachian Mountains.

The most extreme example of thrust structure is in the Alps of Europe. Here, in the various ranges of the Alps, thrust sheets and great overfolds have been moved many kilometers and piled one upon the other, as shown in the cross section (Fig. 3.58).

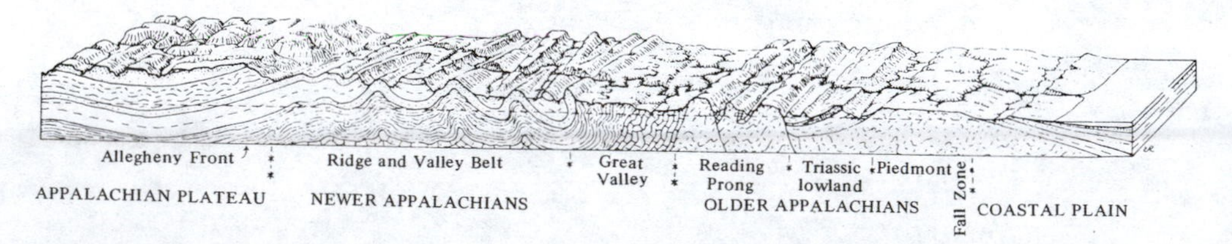

FIGURE 3.56
Generalized cross section of Appalachian Mountains in Pennsylvania, showing the folded sedimentary rocks that form the mountains and the metamorphic and igneous rocks of the piedmont. Gently tilted sedimentary rocks form the coastal plain. From Johnson, 1931.

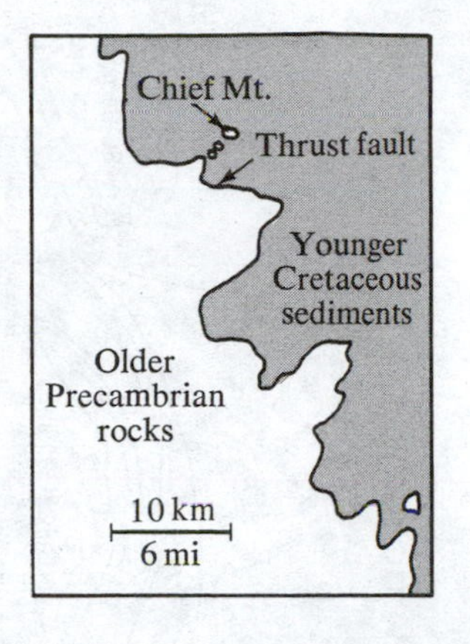

FIGURE 3.57
Map showing Chief Mountain thrust fault, Glacier National Park, Montana. The older Precambrian rocks have been thrust over younger Cretaceous rocks. The minimum displacement on the thrust is the distance from Chief Mountain, which is an erosional remnant of the thrust plate, to the farthest west point where the thrust is exposed. Movement was from west to east. (After C. P. Ross, D. A. Andrews, and I. J. Witkind, *Geologic Map of Montana,* U.S. Geological Survey, 1955)

Other mountain ranges are dominated by metamorphic and granitic rocks. Examples are the eastern part of the Appalachians (the Piedmont area, which has been largely eroded), the Sierra Nevada range of California, and the northern Rocky Mountains of Idaho. In these areas the older sedimentary and volcanic rocks have been largely obliterated by intrusive rocks. The folded, older rocks are preserved at places on the flanks and, more rarely, in the interior of the range.

Most of the structures shown in this section appear to be compressional. This conclusion seems clear, especially in areas where rigid, metamorphic basement rocks have been folded along with the more plastic sedimentary rocks. At other areas, only the near-surface rocks have been compressed; the deeper rocks are not deformed. This is the case with some thrust faults where the upper plate is not strong enough to withstand the pressure necessary to push it. In such cases, perhaps uplift produces a slope down which gravity causes the block to slide. Such a mechanism is mechanically possible on rather gentle slopes. Folding also can be caused by plastic beds slumping downslope. This can be seen, on a small scale, in many sedimentary beds (Figs. 3.59 and 3.60).

Vertical movements on the continents are easy to prove. Marine sedimentary rocks high on mountain ranges certainly prove such movements. Areas below sea level, such as Death Valley in California, imply downward movement. The only agent that could have eroded Death Valley below sea level is the wind, but the wind is much too feeble an agent of erosion to have done this.

Other evidences of vertical movements are the uplifted, wave-cut beaches on many seacoasts, especially along the west coast of the United States. At other places the coastline shows evidences of submergence, which is clearly indicated when river valleys are flooded, making estuaries (Fig. 3.61).

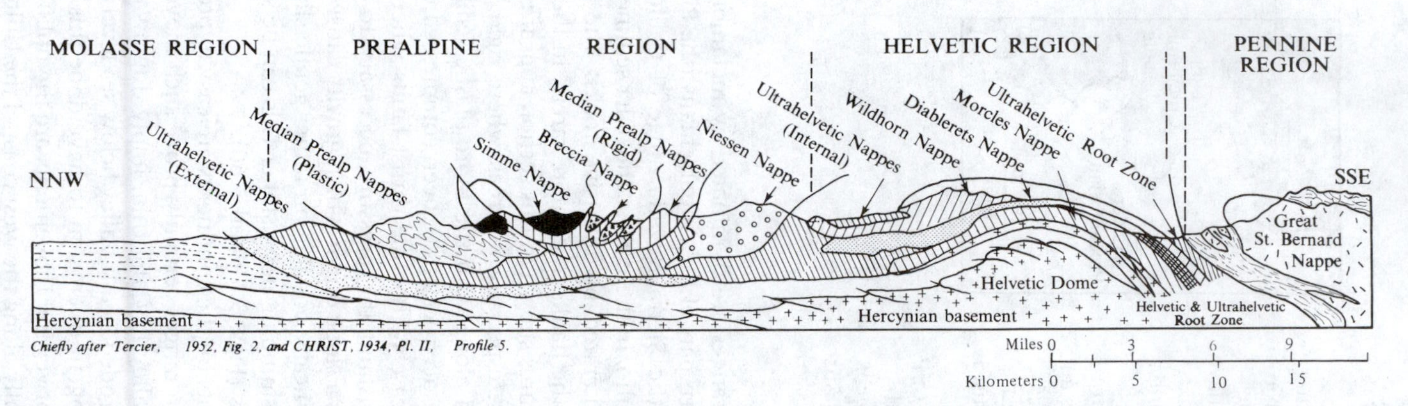

FIGURE 3.58
Cross section through part of the Alps. Nappe is the term applied in the Alps to thrust sheets and great overfolds. Note the large distances that the overthrust rocks have moved. (From Crowell, 1955)

FIGURE 3.59
Gravity thrust fault. The block broke away and slid down the slope.

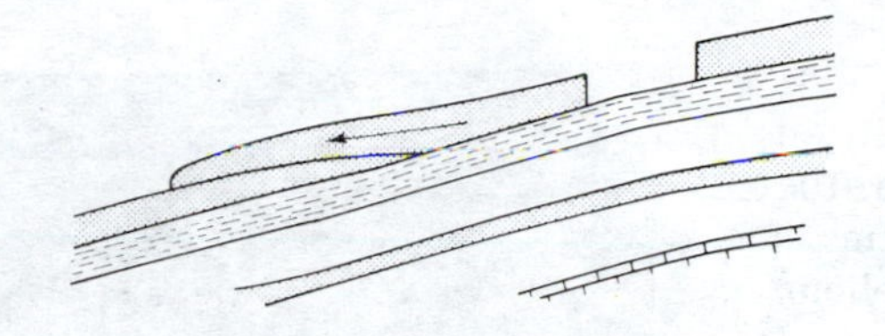

FIGURE 3.60
Gravity folding. Plastic sediments tend to flow downslope, producing folds.

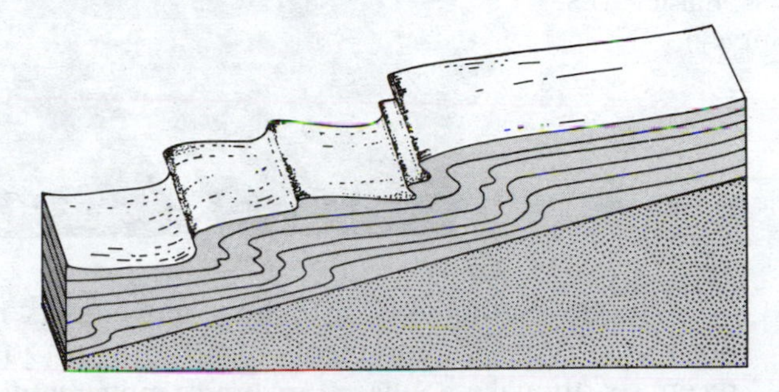

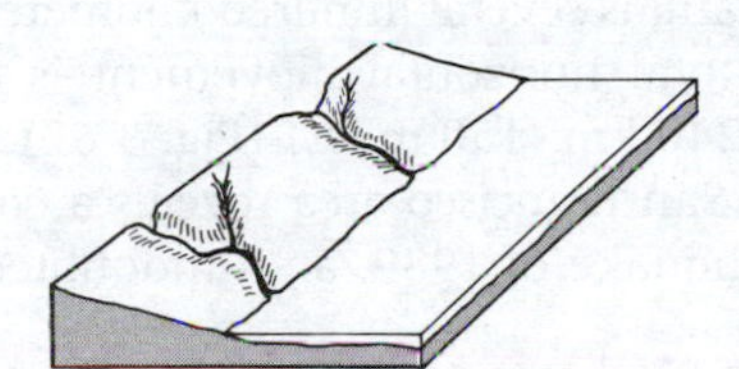

A. Neutral coastline

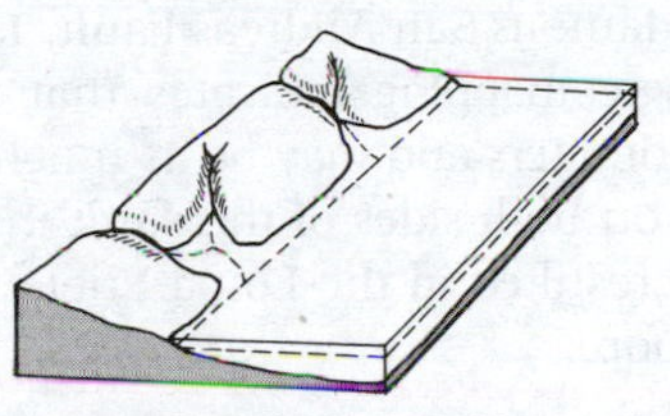

B. Submergent coastline. Wave action may develop cliffs on the headlands, and the eroded material may form sand bars across the bays.

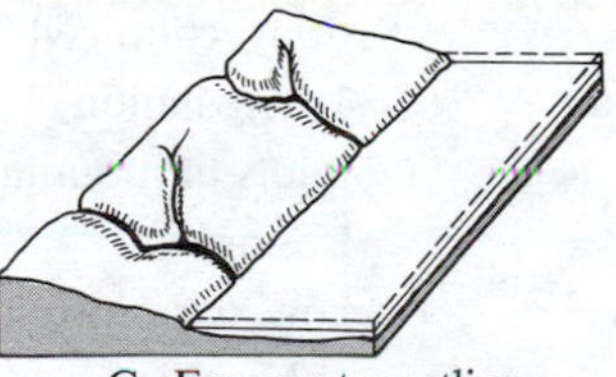

C. Emergent coastline

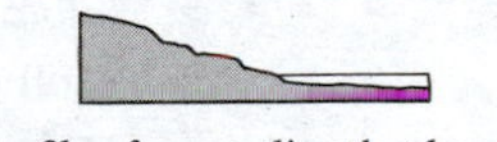

D. Profile of a coastline that has been uplifted (or the sea level fallen) several times, producing a series of wave-cut terraces and cliffs.

FIGURE 3.61
Development of coastlines. A. A neutral coastline. B. The same coastline after the land has sunk or sea level risen. Note the drowning of the river valleys, producing bays and estuaries. Such irregular coastlines are typical of submergence. C. The same coastline after the land has risen (or sea level lowered). Note the wave-cut cliff produced by breaking waves on the newly exposed land. The rivers form falls where they meet the ocean. Offshore sand bars may form. D. The profile of a coastline that has emerged in several stages similar to C. The recent rise of sea level caused by the melting of glaciers has resulted in submergence of most coastlines; however, examples similar to D are common on the west coast of North America.

FIGURE 3.62
Aerial photo of the San Andreas Fault, California. The crushed fault zone forms the diagonal line on the photo. The fault is crossed by a highway in the foreground. (Photo by J. R. Balsley, U.S. Geological Survey)

Recent earthquakes also show movements in the crust, both vertical and horizontal. Vertical movements of up to 6 m (20 feet) have been measured on recent earthquake faults. Horizontal movements of this same amount were measured after the San Francisco earthquake of 1906. The fault involved in this case is the famous San Andreas Fault. This fault is several hundred kilometers long, and geologic mapping indicates that the total horizontal movement is at least several kilometers and may be as much as 240 km (150 miles) (Fig. 3.62). Precise surveying on both sides of this fault in the San Francisco area reveals a continual creep that resulted in the Loma Prieta earthquake of 1989, and undoubtedly will produce more.

KEY TERMS

Seismology
Earthquake
P wave
S wave
Surface wave
Seismograph
Magnitude
Focus
Epicenter
Crust
Mantle
Core
Mohorovičić discontinuity
M-discontinuity
Moho
Meteorite
Ocean basin floor
Submarine mountain
Atoll
Mid-ocean ridge
Continental slope
Continental shelf
Volcanic arc
Negative gravity anomaly
Continental drift
Sea-floor spreading
Magnetic stripe
Plate tectonics
Fracture zone
Transform fault
Hot spot
Mantle plume
Convection cell
Triangulation
Roots of mountains
Isostasy
Low-seismic-velocity zone
Lithosphere
Asthenosphere
Syncline
Anticline
Reverse fault
Thrust fault
Normal fault
Geosyncline
Sedimentation stage
Orogenic phase
Plate collision

QUESTIONS

1. What is the average specific gravity of the earth, and what problem does this pose?
2. How are earthquakes caused?
3. How does a seismograph work?
4. Briefly describe each of the three types of seismic waves.
5. Which of the earthquake waves causes the most damage? How might some of this damage be avoided?
6. What does the magnitude of an earthquake mean?
7. How is the epicenter of an earthquake determined? (A diagram might help.)
8. Describe the behavior of S waves passing through the earth.
9. Describe the behavior of P waves passing through the earth.
10. How is the focus of an earthquake determined?
11. The deepest earthquakes are ______________ km (______________ miles).
12. The core has a radius of ______________ km (______________ miles) and is believed to be composed of ______________.
13. The mantle of the earth is ______________ km (______________ miles) in radial thickness and is believed to be composed of ______________.
14. The crust is separated from the mantle by the ______________.
15. How can earthquakes be predicted?
16. How can people trigger earthquakes?
17. The lithosphere is separated from the asthenosphere by the ______________.
18. Sketch the ocean-continent boundary at a stable place, and label the parts.
19. What does a negative gravity anomaly mean?
20. Describe a typical island arc and sketch a cross section.
21. Describe the similarities between island arcs and geosynclines; between island arcs and continental slopes.
22. Describe the submarine mountains of the Pacific (composition, arrangement, etc.).
23. What evidence suggests that the southern continents were once joined?
24. How do flat-topped seamounts form? What does this suggest about the history of the oceans?
25. How long are the mid-ocean ridges?
26. Describe a cross section of the mid-ocean ridges. (A sketch might help.)
27. What is the evidence for sea-floor spreading?
28. What evidence do you feel is strongest in suggesting continental drift? What evidence is strongest against?
29. How is the magnetism of oceanic rocks used as evidence for continental drift?
30. What are the margins of lithospheric plates? (There are three types of margins.)
31. Describe transform faults.
32. What are lithospheric plates? How many are there?
33. Where do magmas originate?
34. What is the source of the energy that causes deformation of the earth's crust?
35. Prove to a friend who has taken no science courses that crustal movements have occurred.
36. What are convection currents? What is the evidence for them in the mantle? What other force may move plates?
37. What changes have occurred in the earth's magnetic field?
38. What are the main topographic features of the earth's surface, and what is the geologic significance of the two most important features?

FIGURE 3.63

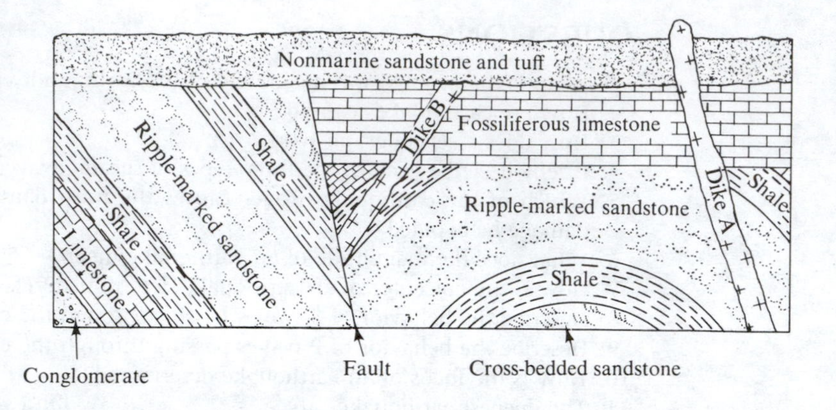

39. The continents are composed of ______________________ .
40. The oceans are underlain by ______________________ .
41. What is the evidence for mountain roots?
42. How is the crust of the earth supported?
43. Is it the strength of rocks that enables mountains to stand above the general level of the continents? Prove it.
44. Why do the continents stand higher than the ocean floors?
45. How is the magnetism of continental rocks used as evidence for continental drift?
46. What type of force produces folds? Normal faults? Reverse faults?
47. What determines whether a rock folds or faults when it is subjected to compression?
48. How could you tell in the field whether a fault is still active?
49. What is an anticline? Discuss several ways one can be distinguished from a syncline.
50. What is a thrust fault?
51. How are the relative ages of the basement rocks forming North America distributed?
52. Outline briefly the history of a geosyncline.
55. Figure 3.63 is a diagram showing a cross section through an area. Starting with the earliest event, list the sequence of events that have occurred in this area.

SUPPLEMENTARY READING

Seismology and Earth's Interior

Anderson, D. L., and A. M. Dziewonski. "Seismic Tomography." *Scientific American* 251(4) (October 1984): 60–68.

Bolt, B. A. *Earthquakes: A Primer.* San Francisco: W. H. Freeman & Co., 1978, 272 pp. (paperback).

Boore, D. M. "The Motion of the Ground in Earthquakes." *Scientific American* 237(6) (December 1977): 68–78.

Johnston, A. C. "A Major Earthquake Zone on the Mississippi." *Scientific American* 246(4) (April 1982): 60–68.

Kerr, R. A. "Quake Prediction by Animals Gaining Respect." *Science* 208(4445) (May 16, 1980): 695–96.

O'Nions, R. K., P. J. Hamilton, and N. M. Evensen. "The Chemical Evolution of the Earth's Mantle." *Scientific American* 242(5) (May 1980): 120–33.

Pasteris, J. D. "Kimberlites: A Look into the Earth's Mantle." *American Scientist* 71(3) (May–June 1983): 282–88.

Stevenson, D. J. "Models of the Earth's Core." *Science* 214(4521) (November 6, 1981): 611–19.

Walker, Jearl. "How to Build a Simple Seismograph to Record Earthquake Waves at Home." *Scientific American* 241(1) (July 1979): 152–61.

Wesson, R. L., and R. E. Wallace. "Predicting the Next Great Earthquake in California." *Scientific American* 252(2) (February 1985): 35–44.

General

Ben-Avraham, Zvi. "The Movement of Continents." *American Scientist* 69(3) (May–June 1981): 291–99.

Bonatti, Enrico, and Kathleen Crane. "Ocean Fracture Zones." *Scientific American* 250(3) (May 1984): 40–49.

Carrigan, C. R., and David Gubbins. "The Source of the Earth's Magnetic Field." *Scientific American* 240(2) (February 1979): 118–30.

Cook, F. A., L. D. Brown, and J. E. Oliver. "The Southern Appalachians and the Growth of Continents." *Scientific American* 243(4) (October 1980): 156–88.

Courtillot, Vincent, and G. E. Vink. "How Continents Break Up." *Scientific American* 249(1) (July 1983): 43–52.

Gass, I. G. "Ophiolites." *Scientific American* 247(2) (August 1982): 122–31.

Hamilton, W. B. "Plate Tectonics and Volcanic Arcs." *Geological Society of America Bulletin* 100(9) (October 1988): 1503–27.

Jordan, T. H. "The Deep Structure of the Continents." *Scientific American* 240(1) (January 1979): 92–107.

Macdonald, K. C., and B. P. Luyendyk. "The Crest of the East Pacific Rise." *Scientific American* 244(5) (May 1981): 100–16.

Marsh, B. D. "Island-Arc Volcanism." *American Scientist* 67(2) (March–April 1979): 161–72.

Molnar, Peter, and Paul Tapponnier. "The Collision between India and Eurasia." *Scientific American* 236(4) (April 1977): 30–41.

Moorbath, Stephen. "The Oldest Rocks and the Growth of Continents." *Scientific American* 236(3) (March 1977): 92–104.

Oliver, Jack. "Exploring the Basement of the North American Continent." *American Scientist* 68(6) (November–December 1980): 676–83.

Pollack, H. N., and D. S. Chapman. "The Flow of Heat from the Earth's Interior." *Scientific American* 237(2) (August 1977): 60–76.

Sclater, J. G., and C. Tapscott. "The History of the Atlantic." *Scientific American* 240(6) (June 1979): 156–75.

Smith, R. B., and R. L. Christiansen. "Yellowstone Park and a Window on the Earth's Interior." *Scientific American* 242(2) (February 1980): 104–17.

Vink, G. E., W. J. Morgan, and P. R. Vogt. "The Earth's Hot Spots." *Scientific American* 252(4) (April 1985): 50–59.

Windley, B. F. *The Evolving Continents.* New York: John Wiley & Sons, 1977, 385 pp.

4

Principles of Historical Geology

Historical geology, the history of the earth, is perhaps the most important goal of geology. In this chapter the methods used to determine geologic history will be described. Some of these methods already have been discussed, such as interpretation of the formation of a rock based on study of the rock itself and its relationship to other rocks. Simple examples are the interpretation of a sandstone composed of well-rounded quartz grains as probably the product of deep weathering and long transport, and recognition that an intrusive rock is younger than the rock it intrudes. Geologic history is determined from interpretation of the way rocks were formed and the dating of these events. This chapter is concerned mainly with methods of dating so that rocks and events can be placed on a time scale.

The actual history of the earth will not be described here because the narrative is too long. In the broadest sense the earth's history is the formation, movement, collision, and consumption of crustal plates. Continental geology is dominated by repeated incursions of the sea and development of geosynclines, all of which appear to be related to the plates. Thus, many aspects of historical geology already have been covered.

RELATIVE AND ABSOLUTE DATES

Rocks can be dated both relatively and absolutely. **Relative dates** are of two types. The simpler of these is used to determine the sequence of events in a limited area. This method uses superposition and cross-cutting relationships. Beds are dated by **superposition** from the knowledge that, in undisturbed sedimentary rocks, the oldest beds are at the bottom of the sequence and the youngest at the top (Fig. 4.1A). **Cross-cutting features**—faults, and intrusions like dikes and batholiths—are younger than the rocks they cut (Fig. 4.1B). This type of relative date is determined from the relationships among the rocks, and these relationships are determined by geologic mapping. Geologic mapping is the main method of geologic study, and consists of plotting the locations of rock units on an accurate map. *The second way of relative dating is by referring rocks to a geologic time scale, based on study of the fossils in the rocks.* **Absolute dating** is a more recent development and *is based on radioactive decay of elements in rocks.* This method has been used to calibrate the relative geologic time scale developed from the study of fossils.

GEOLOGIC TIME SCALE

Recognition of the immensity of geologic time and the development of methods for subdividing geologic time are among our great intellectual accomplishments. Discoveries made near the end of the eighteenth century led to rapid progress, although there had been earlier studies. The culmination of this early work was the

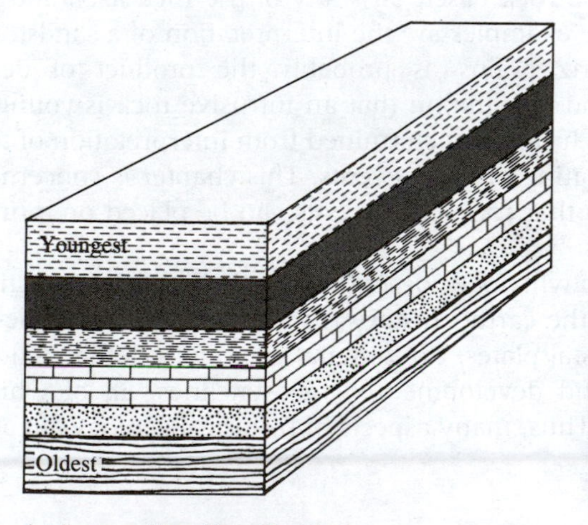

A. *Superposition*. The oldest beds are at the bottom and the youngest at the top.

B. *Cross-cutting*. The batholith is younger than the beds that it intrudes.

FIGURE 4.1
Relative dating.

theory of uniformitarianism, published by James Hutton in 1785 and 1795. Uniformitarianism, as used in geology, means that the present is the key to the past; that is, the history of old rocks can be interpreted by noting how similar rocks are being formed today. Hutton's book was poorly written, and the theory did not gain wide acceptance until it was popularized by John Playfair in 1802. Uniformitarianism is a simple idea, but it is the keystone of modern geology.

A simple example will show the power of this theory. Many varieties of shell-bearing animals can be seen in the mud along the seashore at low tide. A similar mudstone with similar shells encountered in a canyon wall is interpreted, using uniformitarianism, as a former sea bottom that has been lithified and uplifted. If this seems too elementary, we need only remember that just a few years before Hutton's time, few people recognized that fossil shells were evidence of once-living organisms (Fig. 4.2).

After several false starts, the geologic time scale was developed in the last part of the eighteenth century and the first half of the nineteenth. In 1782 the great French chemist Antoine Lavoisier demonstrated that quarries dug for pottery and porcelain clay near Paris all exposed the same sequence of sedimentary rocks. Georges Cuvier and Alexandre Brongniart published maps in 1810 and 1822 showing the distribution of the various rock types around Paris, and in 1815 William Smith published a geologic map of England. Cuvier and Brongniart studied the fossils in each of the sedimentary layers that they mapped, and discovered, as did Smith, that each layer contained a different group of fossils. They discovered

FIGURE 4.2
Miocene fossils from Virginia. Snails and clams are most abundant. (Photo by W. T. Lee, U.S. Geological Survey)

that a sedimentary bed could be identified by its fossils. That work set the scene for the next 20 years, for during this period the geologic time scale was developed.

The **geologic time scale** or column was the result of individual work by many people in western Europe. These individuals studied the rocks and the fossils they contained at well-exposed places, generally near their homes. As the studies proceeded in this haphazard manner, the general sequence of older to younger was recognized and the gaps were filled. It also was shown that rocks in similar stratigraphic position, although far removed from the type area, contained the same or very similar fossils. Thus, it took work by many people spread over a wide area to demonstrate that fossils can be used to date rocks. The geologic time scale in current use is shown in Figure 4.3. Almost all of the **systems** or **periods** were originally defined in the first half of the nineteenth century.

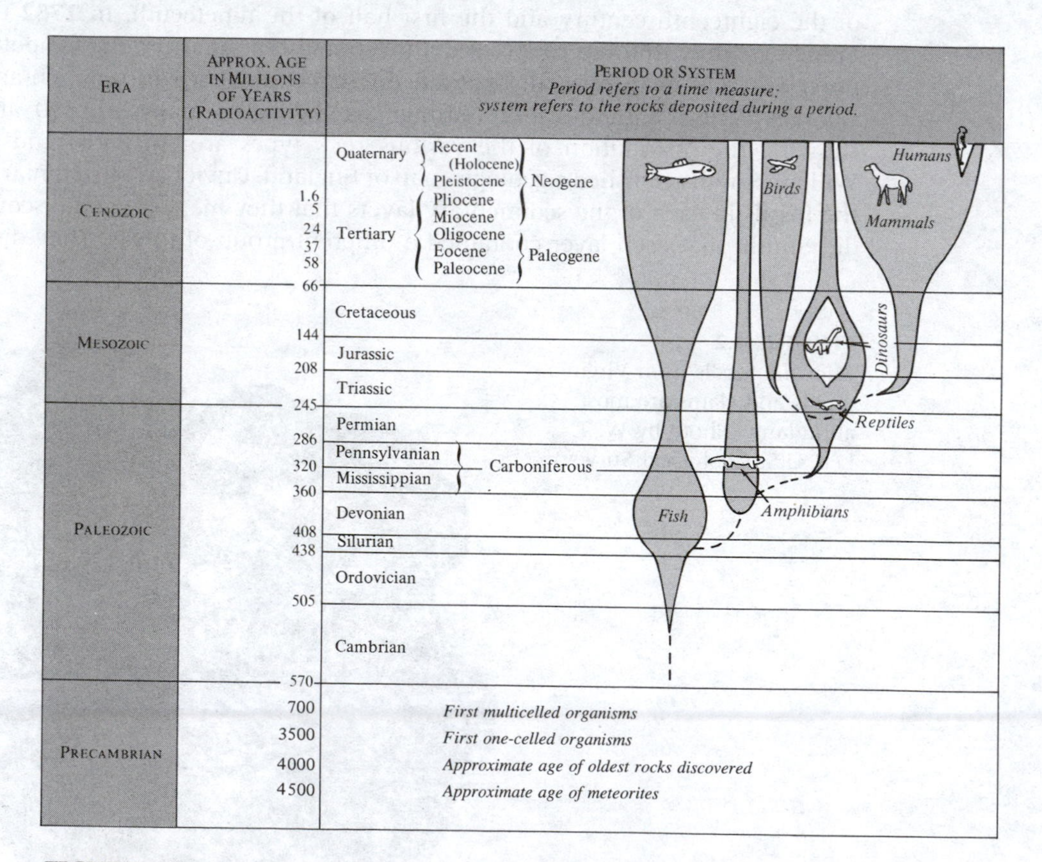

FIGURE 4.3
The geologic time scale. At right is a very simplified diagram showing the development of vertebrates. Not included on the diagram are many types of invertebrate fossils such as clams, brachiopods, corals, sponges, and snails, which first appeared in the Cambrian or Ordovician Period and have continued to the present.

Development of the geologic time scale showed the general development of life on earth, setting the stage for Charles Darwin, who published his theory of evolution in 1859. Although the idea of evolution was not invented by Darwin, his book started a still-continuing controversy. Until about this time no controversy existed between geology and religion because geologists interpreted geologic history in biblical terms. However, in the preceding hundred years some geologists began to recognize that a much longer time was necessary for the deposition of the sedimentary rocks and other geologic events than students of the Bible would allow. Thus, the central point became the age of the earth.

The geologists and evolutionists needed a very old earth. Estimates of the age of the earth based on the time necessary to deposit the rocks of the geologic time scale were made by many geologists in the latter part of the nineteenth century, and, although they varied between 3 and 1584 million years, about 100 million years was an average figure. The lack of precision and the subjective nature of the assumptions made these estimates suspect. Lord Kelvin, one of the most influential physicists of the day, entered the fray and attempted to calculate the age of the earth from thermodynamics. He assumed that the earth began as a melted body and has been cooling ever since. He calculated that 20–40 million years had passed from the time the earth had cooled enough for life to exist to the present. About the turn of the century, when Kelvin was an old man but still refining his calculations, radioactivity was discovered. As noted earlier, radioactivity is a source of heat within the earth, so Kelvin's heat flow calculations were based on an incorrect assumption. Just a few years after this time, the first attempts were made to use radioactivity to determine the age of the earth; as described later, this ultimately led to the presently accepted age of 4700 million (4.7 billion) years.

UNCONFORMITIES

Once the time scale was set up, geologists traced the geologic systems into previously unmapped areas. One problem that soon appeared was that the systems were originally defined, in many cases, as the rocks between two natural breaks in the geologic record. In other areas there were no breaks, but continuous deposition. Thus, between the original systems were intervals not included in either system, and arguments about the assignment of the neglected intervals are, in some cases, still in progress. The breaks are of several types, such as change from marine to continental deposition, influx of volcanic rocks, and unconformities (Fig. 4.4). This last item is the most important and requires discussion.

Unconformities *are locally unrecorded intervals, or breaks in the geologic record.* They may take several forms. If sedimentary rocks overlie metamorphic rocks, clearly the sediments were deposited after the metamorphism; and because no sedimentary rocks representing the time of metamorphism are present, this is a type of unconformity. In this case, if the metamorphic rocks were formed at considerable depth in the crust, as is generally the case, then the period of erosion that uncovered the metamorphic rocks prior to deposition of the sediments also is missing from the sedimentary record.

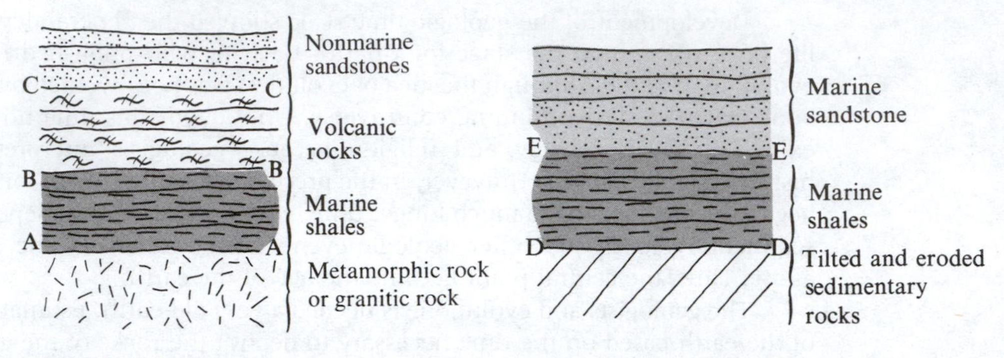

FIGURE 4.4
Natural breaks in the stratigraphic record. A–A and D–D are unconformities because the time necessary for erosion is not represented by sedimentary rocks. The other breaks may or may not be unconformities, depending on whether there is a time interval between the different rock types or continuous deposition. D–D is an angular unconformity.

Another type of unconformity may consist of a change in fossils, representing a short or a long period of time between two beds in the sedimentary section. Such an unconformity may be due to erosion of previously deposited beds or to nondeposition, and may not be at all conspicuous until fossils are studied.

The third type of unconformity is more obvious and consists of folded and eroded sedimentary rocks that are overlain by more sedimentary rocks. Such an unconformity is called an **angular unconformity** *because the bedding of the rocks above and below it are not parallel.* In this, the time of folding and erosion is not represented by sedimentary rocks.

CORRELATION AND DATING OF ROCKS

So far we have considered how the geologic systems were defined, and it is obvious that they are recognized in previously unstudied areas by comparing the contained fossils with the fossils from originally defined type areas. In this manner the systems defined in Europe were recognized in North America and the other continents. The type of **correlation** by fossils also is most important in establishing the equivalence of nearby beds. Other methods of correlating nearby beds include actually tracing them from one area to another by walking out the beds. In other cases distinctive materials in a bed may be used to identify it, such as fragments of an unusual rock type or a distinctive or uncommon mineral or assemblage of minerals. Another method is to use the sequence of beds (Figs. 4.5 and 4.6).

Note that the methods of correlation, with the exception of using fossils, only establish the *continuity* of a sedimentary bed. Fossils establish the fact that both beds are the *same age.* The difference between these two types of correlation can be seen by considering an expanding sea, a not uncommon occurrence in geologic history.

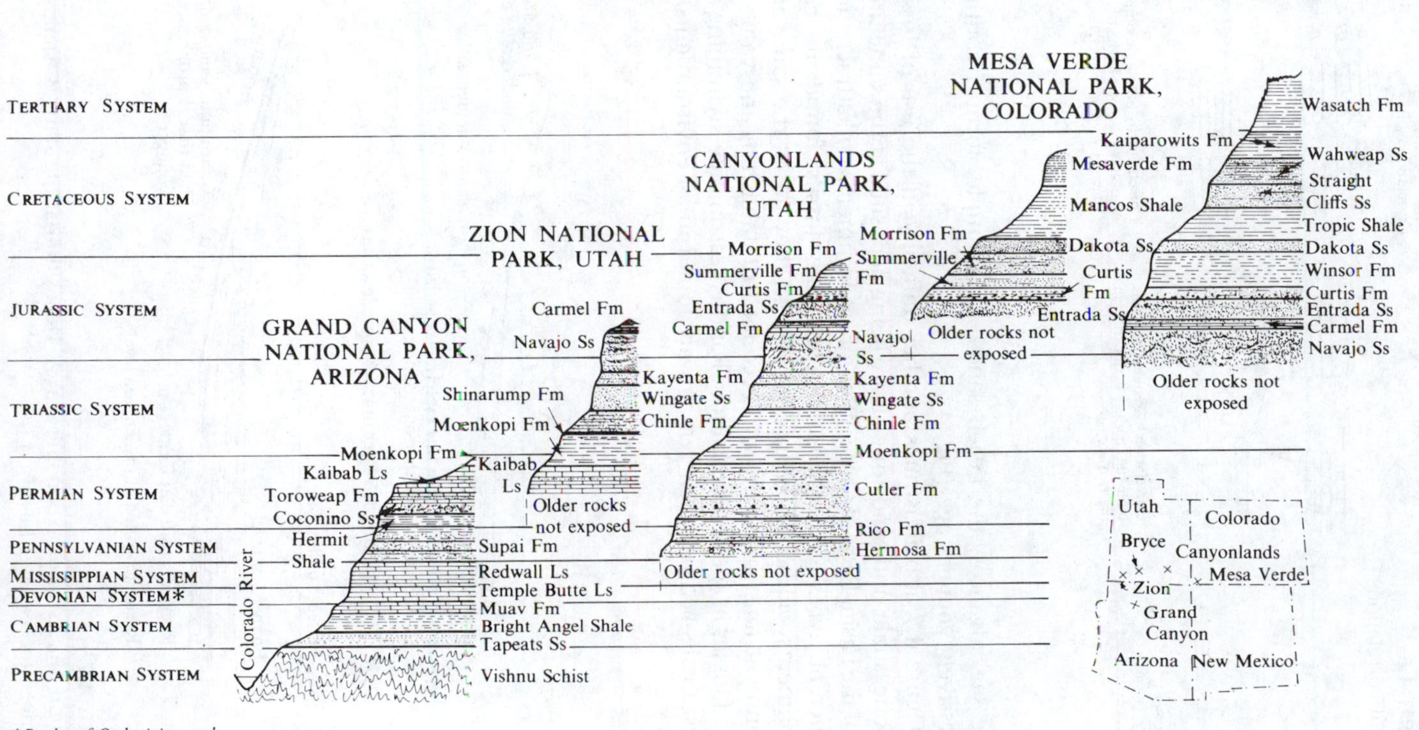

*Rocks of Ordovician and Silurian age are not present in the Grand Canyon.

FIGURE 4.5
Correlation of the strata at several places on the Colorado plateau reveals the total extent of the sedimentary rocks. (After U.S. Geological Survey)

FIGURE 4.6
Types of correlation.

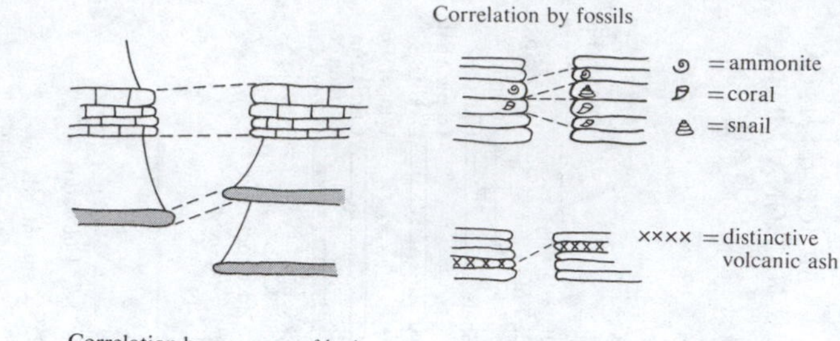

Correlation by sequence of beds. Note the changes in thickness in the two areas.

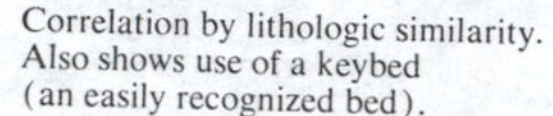

Correlation by lithologic similarity. Also shows use of a keybed (an easily recognized bed).

As the sea advances, the beach sands, for example, form a continuous bed of progressively younger age (Fig. 4.7).

Figure 4.7 also illustrates another difficulty in the use of fossils. Note that limestone, shale, and sandstone all are being deposited at the same time in different parts of the basin. Each of these different environments will attract and be the home of different types of organisms. Thus, the fossils found in each of these environments will be different, even though they are of the same age. Therefore, environment must be taken into account in establishing the value of a fossil or group of fossils in age determination. For this reason, free-swimming animals make the best fossils because their remains are found in all environments.

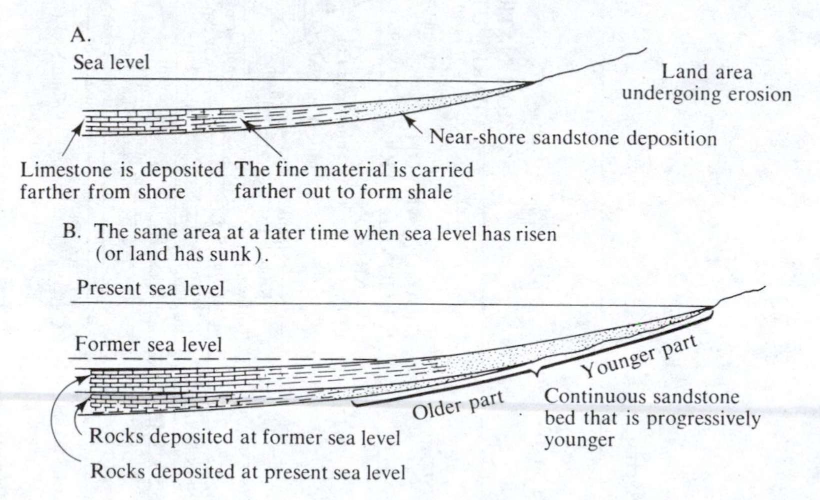

FIGURE 4.7
Deposition in a rising or expanding sea. A. Deposition at the start. B. Deposition at a later time, after sea level has risen.

Fossils that clearly establish the age of the enclosing rocks are called **index fossils.** Ideally, an index fossil should be widespread in all environments, be abundant, and have existed over a short time span. Very few fossils meet all of these requirements, so generally groups of fossils are used to establish the age of a bed (Fig. 4.8).

The only rocks that have reasonably abundant fossils are marine sedimentary rocks, and not even all of these have enough fossils to establish their age clearly. In the previous pages, we discussed dating of fossiliferous marine rocks. Similar techniques can be used to date continental sediments, but here the number of fossils is generally much smaller. This fact is easy to understand if one compares the number of easily fossilized animals exposed at low tide at the seashore with the much smaller number of animals living in a similar-sized area of forest or grassland. Also consider how easy it is for a clam shell to be buried and so preserved, especially when compared to the chance that a plant leaf or a rabbit skeleton will be buried and preserved as a fossil. A land organism, even if buried, may not be preserved, as oxidizing conditions that destroy organic material exist even at some depth. Also, a dead rabbit is apt to be eaten and dismembered by scavengers. Another problem in the use of continental fossils is that they are controlled even more by climate than are marine organisms. The progression of types of plant and animal life seen during a mountain climb will illustrate this point.

Dating igneous and metamorphic rocks presents even more problems because such rocks almost never contain fossils. They must be dated by their relationships to fossil-bearing rocks. Volcanic rocks are dated by the fossiliferous rocks above and below and, in some cases, by interbedded sedimentary rocks. Intrusives must be younger than the rocks that they intrude; they cannot be dated closer unless the intrusive has been uncovered by erosion and then unconformably covered by younger rocks (Fig. 4.9). Metamorphic rocks are even more difficult to date because both the age of the original rock and the date of the metamorphism should be obtained. The age of the parent rock can be found by tracing the metamorphic rock to a place where fossils are preserved. Finding the date of metamorphism may be more difficult because in general an unconformable cover, if present, gives only the upper limit.

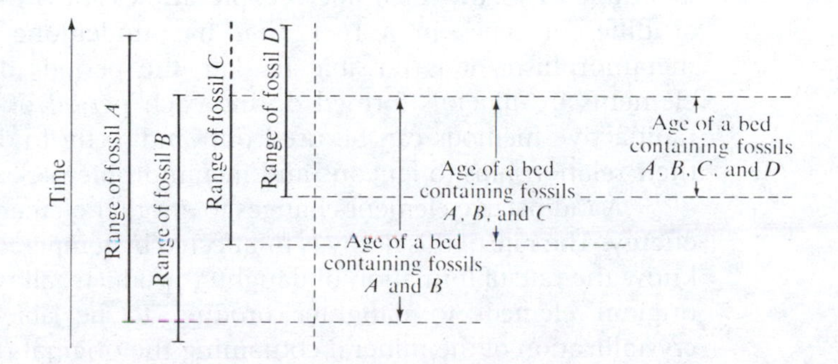

FIGURE 4.8
The use of overlapping ranges of fossils to date rocks more precisely than can be accomplished by the use of a single fossil.

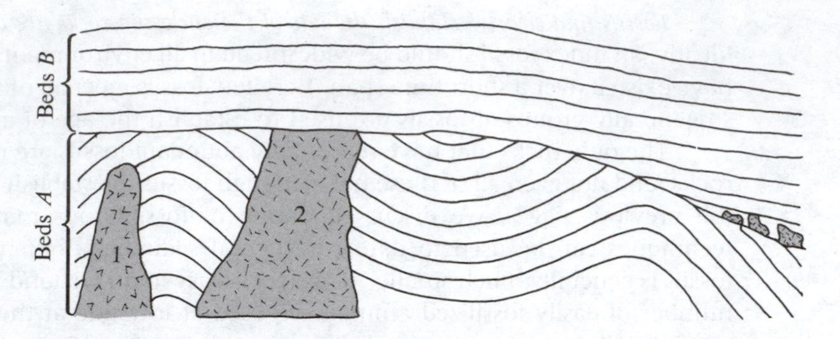

FIGURE 4.9
Dating of intrusive rocks. Intrusive body (1) is younger than beds *A;* no upper limit can be determined. Intrusive body (2) is younger than beds *A* and older than beds *B.* Closer dating may be possible if the conglomerate composed of pebbles of (2) in the basin at the right contains fossils. The contact between beds *A* and *B* is an angular unconformity.

Radioactive Dating

The problems of dating igneous and metamorphic rocks can be overcome by modern radioactive methods. Such methods produce an absolute age in years and so have been used, in addition, to date the geologic time scale in years. In spite of the seeming precision of radioactive dating, it is not yet possible to date a rock as accurately this way as with fossils because of inherent inaccuracies in measurement of the amounts of the elements produced by radioactive decay.

These inaccuracies produce an uncertainty in the radioactive date that is more than the span of zones based on fossils, especially in those parts of the geologic column that contain abundant short-lived fossils.

Radioactive dates are obtained by studying the daughter elements produced by decay of a radioactive element in a mineral. Sometimes more than one element is studied in an effort to provide a check on accuracy. Such a date tells the time of formation of that mineral, and so tells the time of crystallization of an igneous rock or the time of metamorphic recrystallization of a metamorphic rock. In some favorable cases, in which microscopic studies show that different minerals formed at different times in a rock that has undergone more than one period of metamorphism, it is possible to date the periods of metamorphism by dating elements in minerals formed during each period of metamorphism. In general, radioactive methods can be used only indirectly to date sedimentary rocks from their relationships to igneous and metamorphic rocks.

A radioactive element changes to another element by spontaneously emitting energy. The rate of this decay is unaffected by temperature or pressure; hence, if we know the rate of formation of daughter products, all we need to find is the ratio of original element to daughter product to be able to calculate the time of crystallization of the mineral containing the original radioactive element. The rate of decay of a radioactive element is called its **half-life**—*the time it takes for one-half of the original element to decay to daughter products.*

Several radioactive elements are used to date rocks. A few minerals contain uranium, and these minerals generally also contain thorium, which has about the same atomic size as uranium. These minerals are rare and are almost completely confined to pegmatite veins in batholiths. Because such veins are probably the last to crystallize, these occurrences will give only the upper limit of the age of the batholith. These minerals contain thorium and the two radioactive isotopes of uranium. Each of these three produces its own series of radioactive daughter products; each member of each series decays to the next daughter until, finally, a stable isotope of lead is produced. These many steps constitute a possible source of inaccuracy, as it is possible for any one of the daughters to be removed by leaching or some other process. The elements and their final products are:

$$^{238}U \rightarrow {}^{206}Pb$$
$$^{235}U \rightarrow {}^{207}Pb$$
$$^{232}Th \rightarrow {}^{208}Pb$$

Thus, in this case, it is possible to measure three ratios which should all agree in age.

Rubidium is a radioactive element that is generally present in small amounts in any mineral that contains potassium, because the atoms of both elements are about the same size. Rubidium decays in a single step to strontium ($^{87}Rb \rightarrow {}^{87}Sr$). Because potassium minerals that contain rubidium are so common (feldspar and mica) and only a single radioactive decay is involved, this may become the most useful method of radioactive dating.

Potassium also has a radioactive isotope that is useful in dating, especially because it is such a common element. The small amount of radioactive potassium decays to form two daughters:

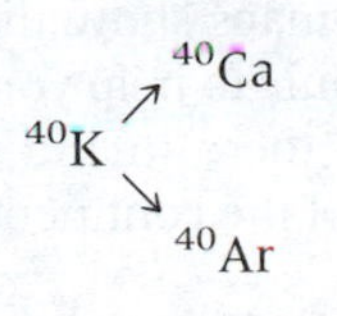

Because the argon is easier to measure, the method used is to obtain the ratio of potassium-40 to argon-40. The main problem is finding minerals that retain the argon, because it is a gas. It appears that micas, in spite of their cleavage, retain most of the argon, but feldspars lose about one-fourth of their argon.

The methods outlined above are useful in the range of about one million to a few billion years. Another method employing the radioactive isotope of carbon is effective in dating carbonaceous material, such as wood or coal, that is less than 40 to 50 thousand years old. This method depends upon the fact that all living matter contains a fixed ratio of ordinary carbon-12 and radioactive carbon-14. Cosmic rays react with nitrogen in the atmosphere to form the carbon-14, which ultimately gets into all living matter via the food cycle. The carbon-14 decays back to nitrogen.

Therefore, the ratio of carbon-12 to carbon-14 can be used to calculate how long an organism has been dead, because a dead organism takes in no more carbon-14. Extensive use of this method is made in archaeology and in glacial studies.

INTERPRETING GEOLOGIC HISTORY

Deciphering the history of the earth or any part of it is a difficult reiterative process with many steps, each step affecting every other step. Multiple working hypotheses must be used to test each step. The process begins with study of the rocks themselves and their relationships to other rocks. This leads to some ideas about their mode of origin. The rocks must be dated and correlated, using the methods described in this chapter. If the rocks have been deformed by folding and faulting, their original relationships and positions must be determined. Each layer or rock unit can then be reconstructed to show the geography at its time of formation. Remember that the rock record is incomplete because of erosion and because many rock layers are exposed at only a few places. The final step is to reconstruct the events that caused the changes between each of the steps.

This brief outline shows how the incomplete rock record is interpreted. Note the various levels of abstraction that are involved. It is no wonder that the history of the earth is not completely known. Interpreting the history of the earth or of a small area is difficult, and revision is constantly necessary as more data are obtained. Thus, geology is an ongoing process that offers great challenge; many fundamental problems remain to be solved.

GEOLOGICAL HISTORY OF MIDDLE NORTH AMERICA

On the following pages is a brief history of part of North America, told in maps and cross sections. The maps show the past environments superposed on the present geography. This is only to help you locate the features; the rivers and other features of today were not there in the geologic past. The world maps that show the probable locations of the continents in the past are subject to revision as more data become available.

Geologic history begins with the origin of the earth. The age of the earth itself is a geological problem to which radioactive dating has been applied. So far, *the oldest rocks found are about 4 billion (4000 million) years old.* This, then, is the youngest that the earth can possibly be, and is probably about the time the crust formed; the earth itself is probably older. Meteorites were discussed earlier and are thought to be the material from which the solar system was formed. Radioactive dates from meteorites give ages of around 4.7 billion (4700 million) years. This suggests that *the earth may have formed 4.5 to 5 billion (4500 to 5000 million) years ago.*

The Precambrian is the oldest era. Most geologic dating is done with fossils, but they do not become abundant and so useful until the Cambrian Period. Because of the resulting poor dating we know only the broadest outline of Precambrian history, even though it makes up almost 90 percent of geologic time. This was the exciting time that the first crust formed, when the ocean and atmosphere formed, and when the first life originated.

In the Hadean Era, the

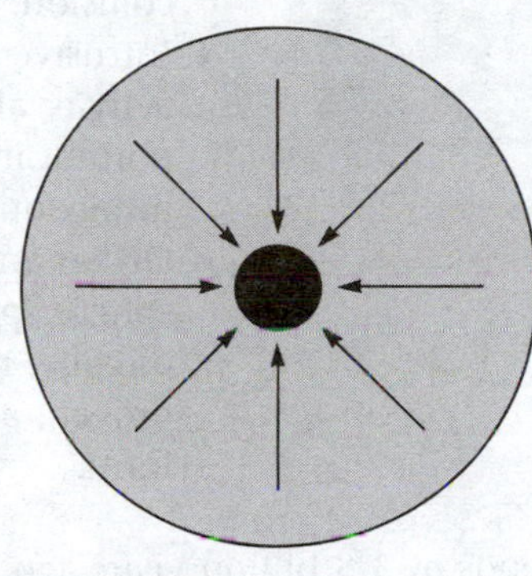

- Initial homogeneous earth melts because of compression and radioactivity.
- Heavy elements move toward the core.
- Less-dense remaining material forms the mantle.

FIGURE 4.10

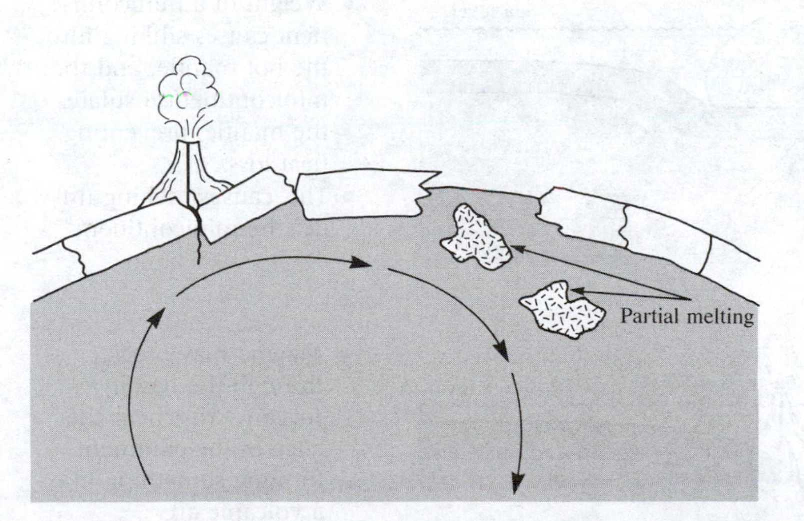

- Mantle cools at the surface and crystallizes, forming the earliest crust.
- Convection in the mantle breaks up the earliest crust and carries it deeper, where the temperature is higher.
- Earliest crust is partially melted, forming basalt and perhaps granite (granite may come from partial melting of basalt or other rocks).
- Meteorite impacts also may break up the earliest crust.

FIGURE 4.11

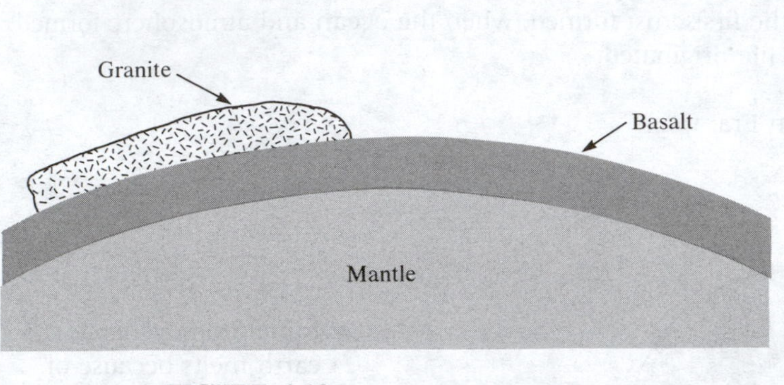

FIGURE 4.12

- Crust forms of basalt, overlain in places by less dense granite. (None of these rocks have been found.)
- Mantle convection moves the crust on the surface so that minicontinents are formed by collision.
- Intrusive and volcanic activity also may be important in forming the minicontinents.
- Ocean and early atmosphere (perhaps carbon dioxide, nitrogen, hydrogen, and water) form.

Archean Era begins and Hadean Era ends at 3.8 billion years ago, with the oldest rocks preserved.

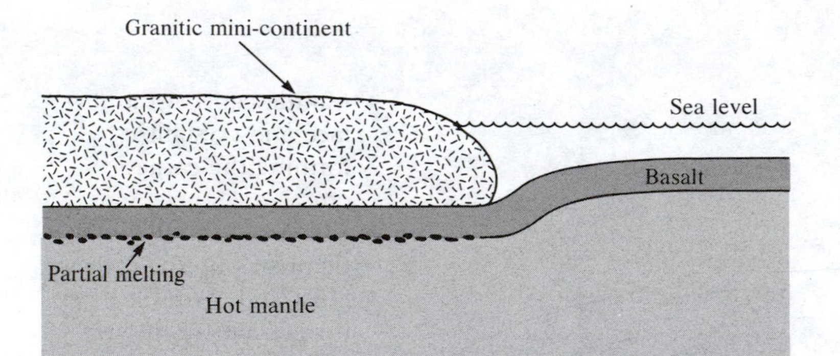

FIGURE 4.13

- Weight of a minicontinent causes sinking into the hot mantle, and the minicontinent insulates the mantle, preventing heat loss.
- This causes melting under the minicontinent.

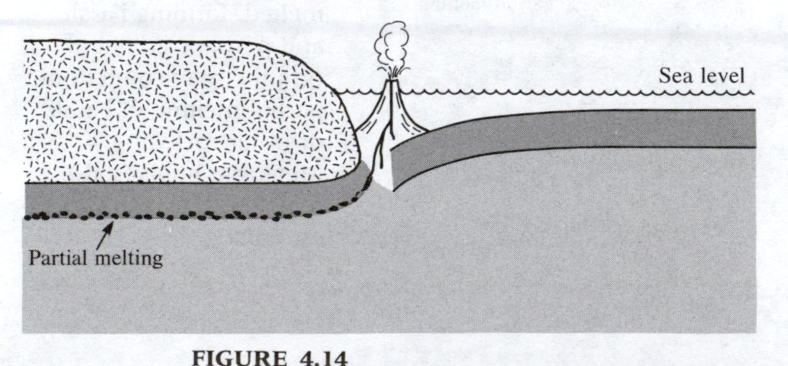

FIGURE 4.14

- Magma may break through the basalt (ocean) crust near the edge of the continent, forming something like a volcanic arc.
- Sediment from erosion of the continent also may accumulate in this area.

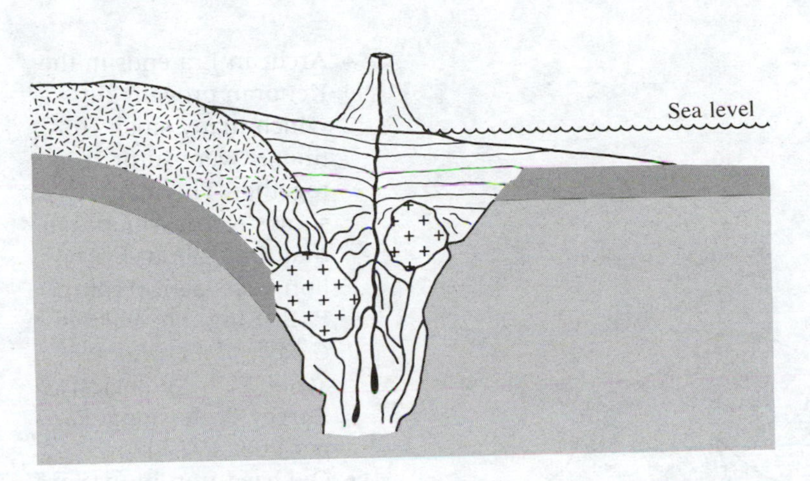

FIGURE 4.15

- The weight of the volcanic and sedimentary rocks may cause sinking, and the resulting melting may increase the mainly basaltic volcanic activity.
- The increased temperature and pressure near the bottom of the downsinking pile causes metamorphism.
- The volcanic and sedimentary rocks become greenstone and the granite becomes high-grade gneiss. Partial melting may produce batholiths. Thus typical Archean rocks are formed.

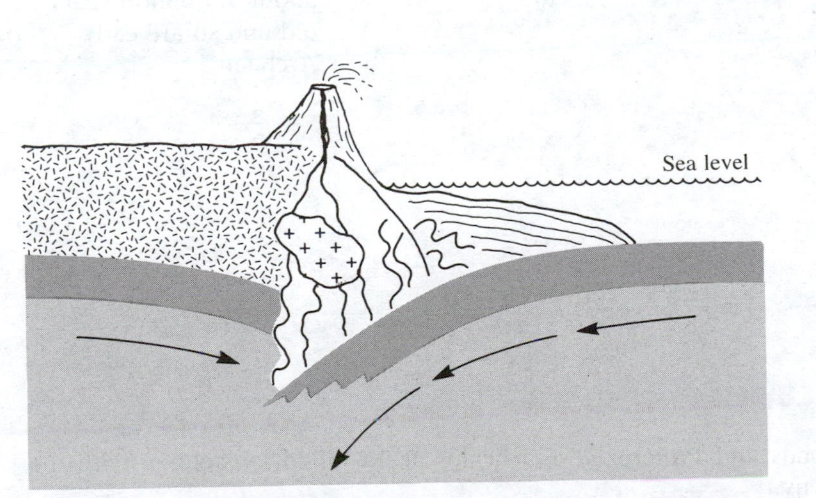

FIGURE 4.16

- Downward movement of the oceanic crust at continent-ocean borders caused by convection in the mantle may form volcanic arcs, thus initiating plate tectonics.
- Igneous and metamorphic activity at the volcanic arcs increases the size of the continents.

- Archean Era ends in the Kenoran orogeny, in which many small continents were sutured together. The map shows North America at that time, about 2.5 billion years ago. (From P. B. King, *Precambrian Geology of the United States,* U.S. Geological Survey Professional Paper 902, 1977.)
- Oxidized iron in marine sedimentary rocks (banded iron formation) shows that oxygen was present in the oceans. Organisms capable of photosynthesis were the probable source of the oxygen.
- The oldest fossils are about 3.5 billion years old and so are early Archean.

FIGURE 4.17

Archean Era ends and Proterozoic Era begins at 2.5 billion years ago with the Kenoran orogeny.

In the early Proterozoic

- Larger sedimentary and volcanic basins formed on the now-more-stable crust.
- Emplacement of many large intrusive bodies at many places suggests the buildup of heat under the larger continents on the still young, energetic earth.
- From 2 billion years on, red beds are common, showing that oxygen was present in the atmosphere.

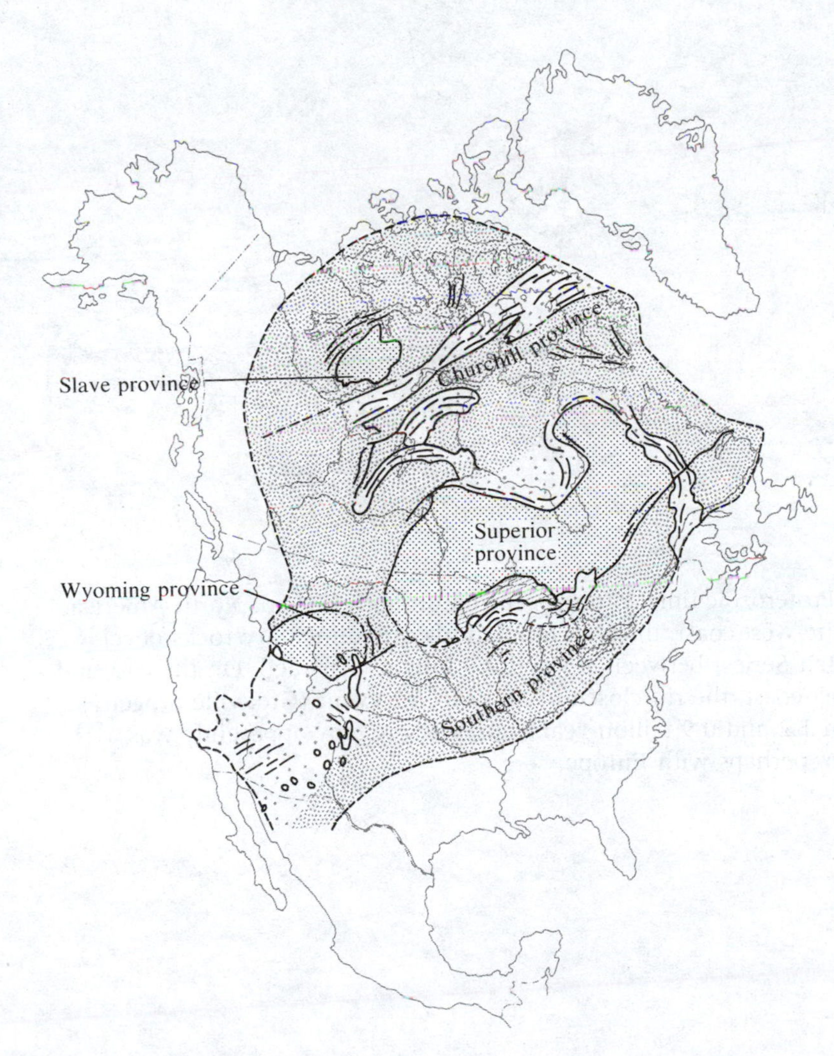

FIGURE 4.18

- About 1.8 billion years ago the core of North America was assembled in the Hudsonian orogeny. (From P. B. King, U. S. G. S. Professional Paper 902.)
- Most of the continents came together at this time to form a supercontinent.

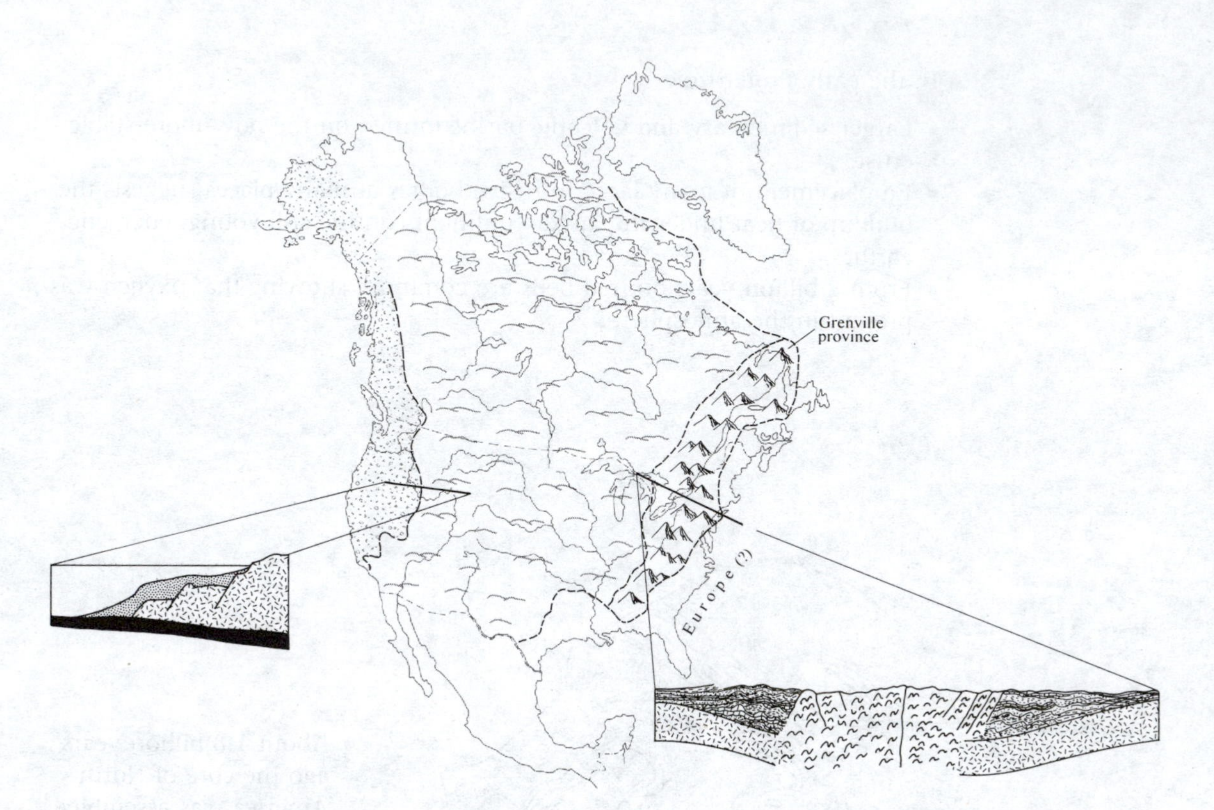

FIGURE 4.19

- In late Proterozoic time, rifting occurred on both sides of North America. Along the west coast, thick accumulations of sedimentary rocks developed (Belt Series) between 1.5 and 0.9 billion years ago. On the east and southeast coast, the rift closed in an orogenic event (Grenville orogeny) between 1.2 and 0.9 billion years ago. The orogeny apparently was a collision, perhaps with Europe.

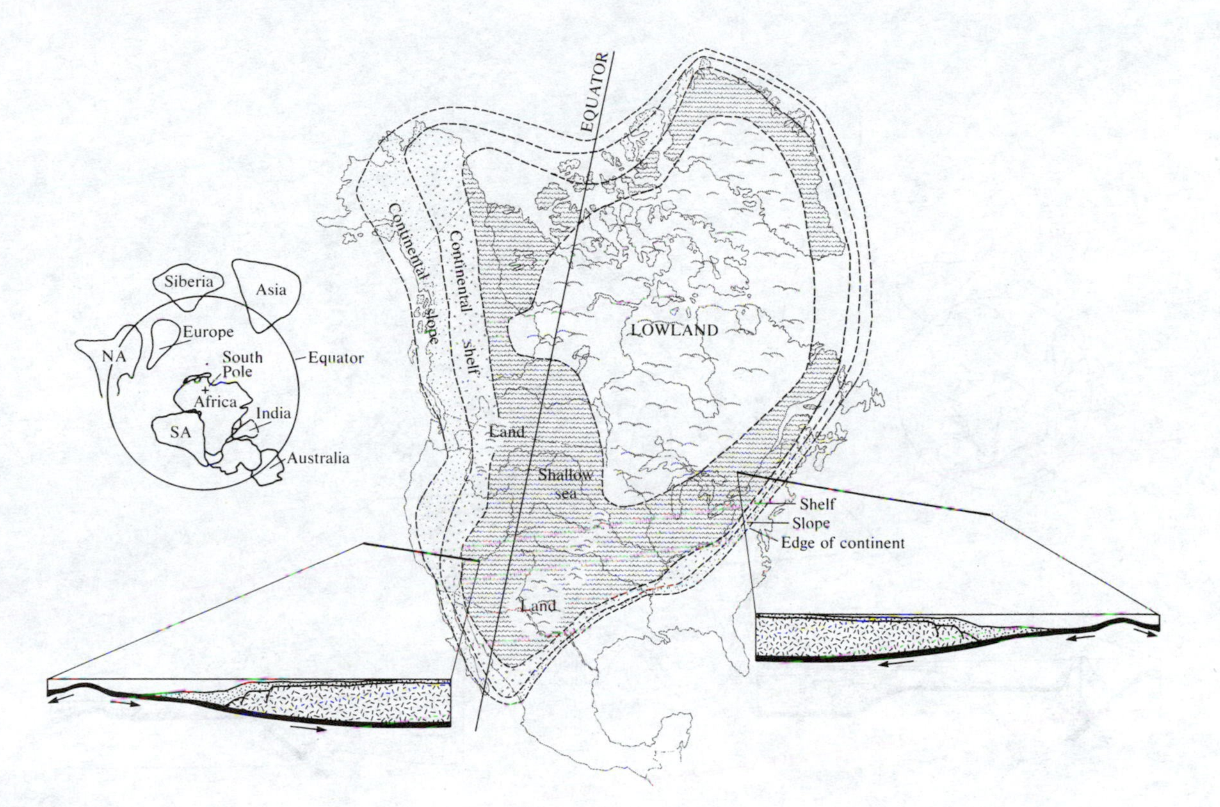

FIGURE 4.20

- Cambrian Period—North America rifted from Europe on the east and perhaps Siberia on the west in late Precambrian. Clastic sediments eroded from the interior were deposited on continental shelf-slope geosynclines at the edge of the continent. (The geography of the present continent is shown for reference only.)

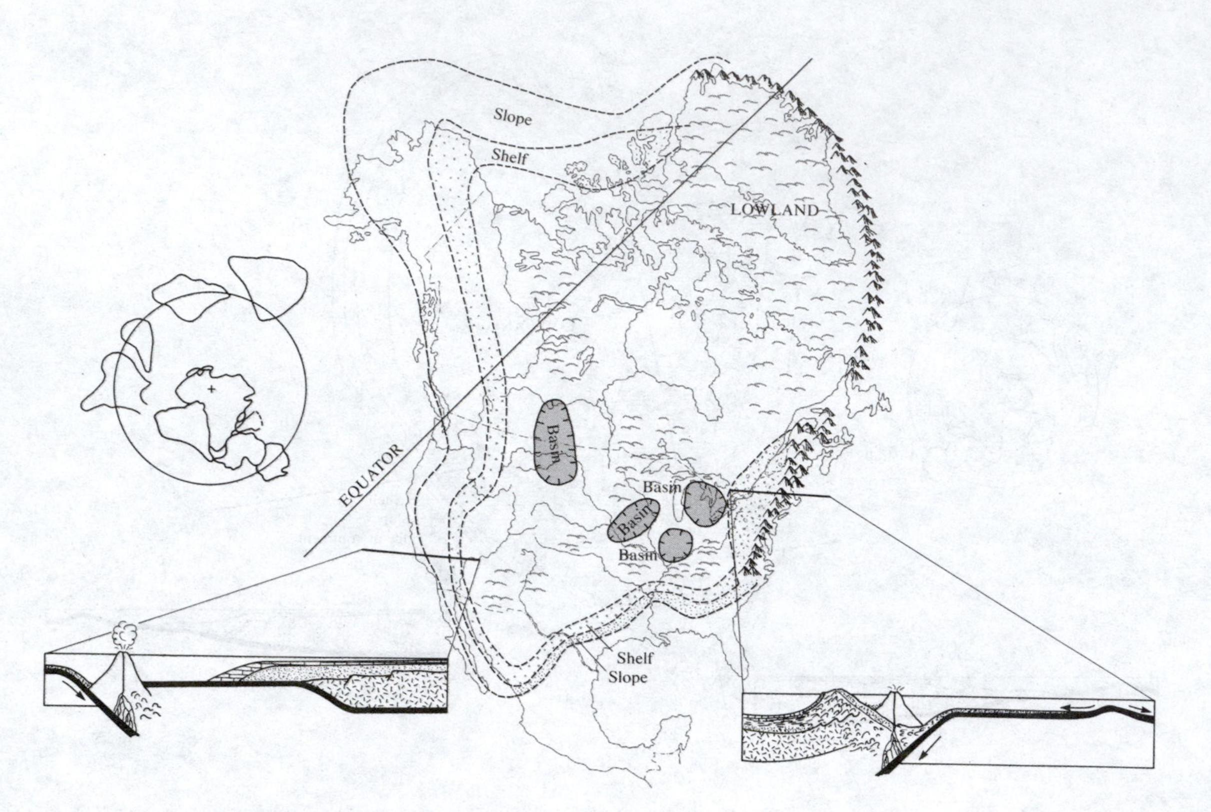

FIGURE 4.21

- In the last half of the Ordovician Period, a volcanic arc formed along the east coast and collided with North America. This deformed the older rocks and created a highland that was the source for the clastic fan. This event is called the Taconic orogeny; it caused thrust faulting in New England. Along the west coast, carbonates accumulated on the shelf and slope, suggesting that erosion had reduced the continental source areas. A volcanic arc was offshore.

These features continued through the Silurian Period.

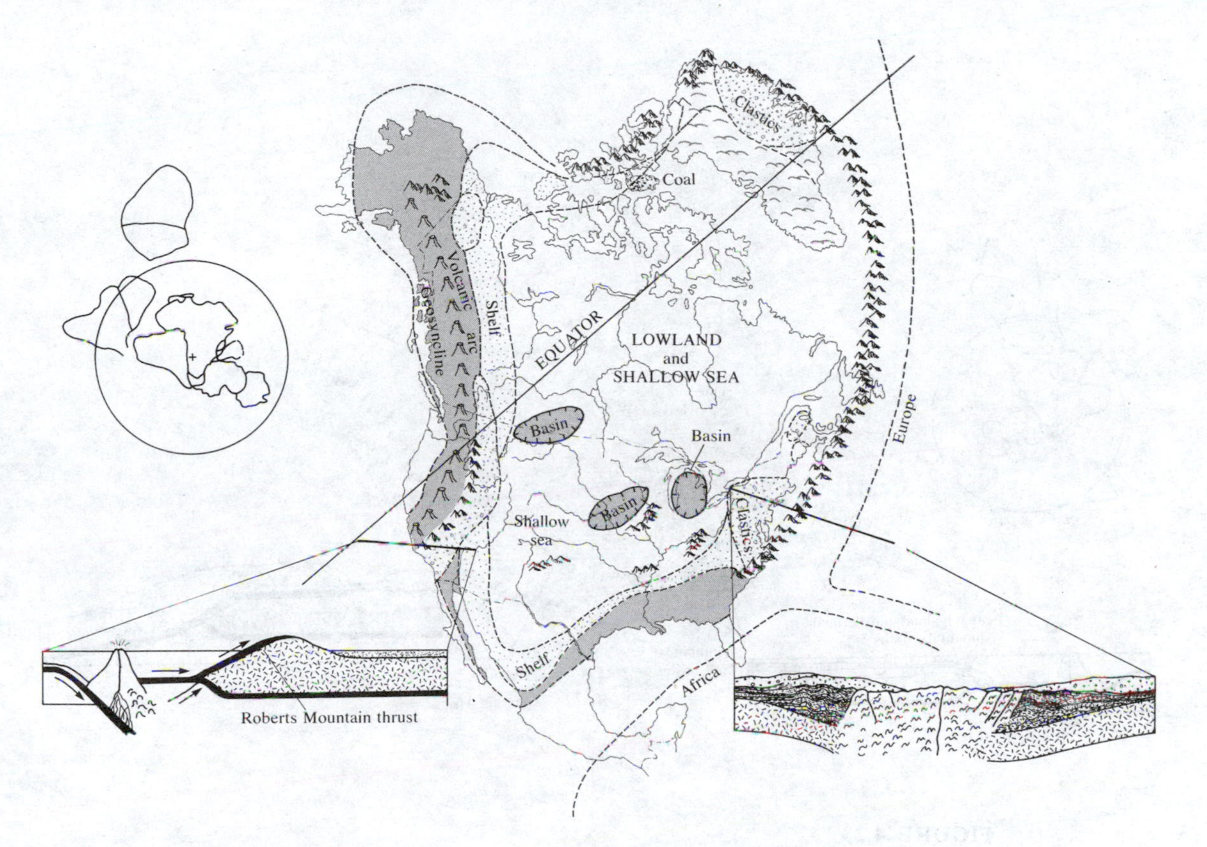

FIGURE 4.22

- In the Devonian Period, collisions occurred on both sides of the continent. About mid-Devonian, Europe collided with the east coast (Acadian orogeny), and clastic fans from the resulting mountains were shed onto both continents. In the west the collision was with a volcanic arc and occurred in late Devonian and Mississippian time (Antler orogeny). Rocks in the basin between the arc and the continent were thrust onto the continent, forming an upland. A clastic fan was deposited to the east of the upland.

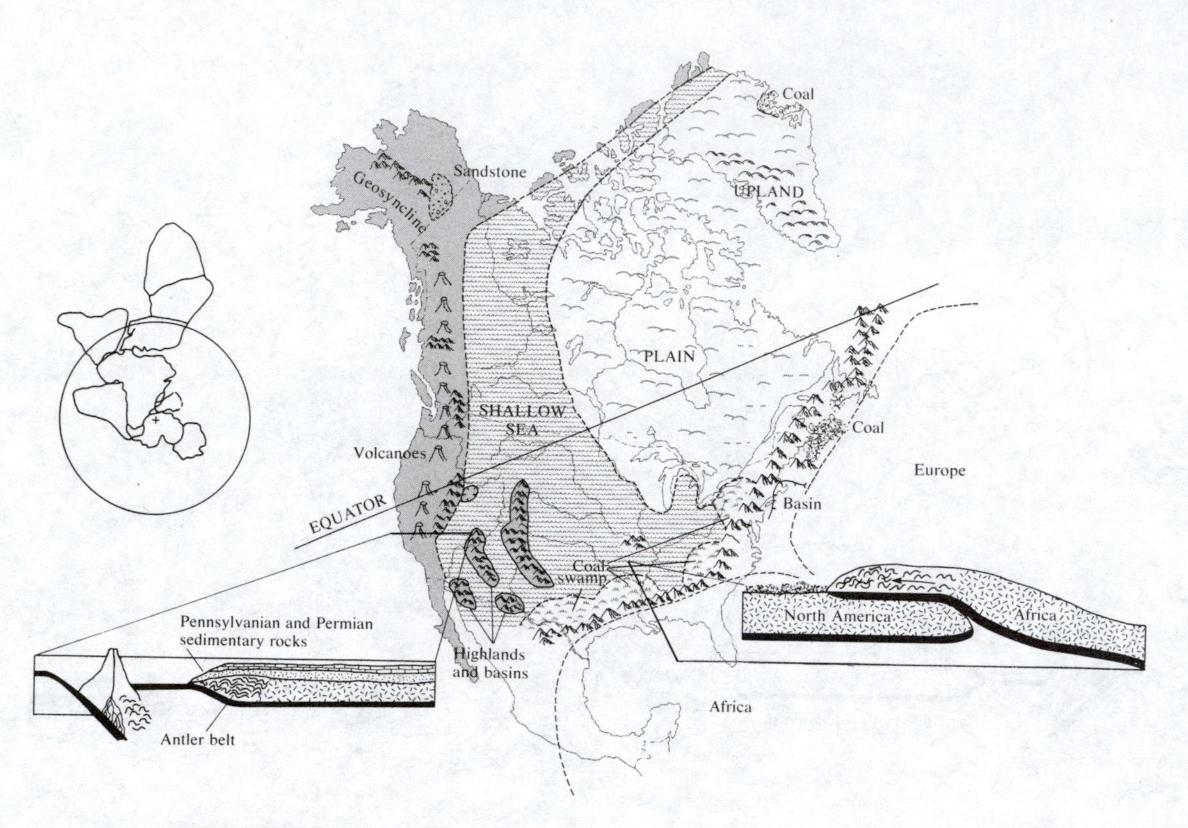

FIGURE 4.23

- In the Pennsylvanian Period, Europe remained in collision with North America and, in Mississippian and Pennsylvanian time, Africa collided with southeastern North America. Coal swamps formed between the inland seas and the uplands produced by the collisions. In the west, sediments were deposited over the area of the Antler orogeny. In Permian time, the seas and swamps dried out in the east. In the west, clastics were deposited near the interior and limestone farther west.

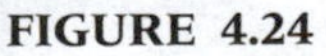

FIGURE 4.24

- In the Triassic Period, the Atlantic Ocean opened as rifting occurred between North America and Europe. This rifting formed basins in late Triassic or early Jurassic time along much of the east coast. On the west coast in early Triassic time, the Sonoma orogeny caused the Golconda thrust in the area west of the Antler orogeny. By late Triassic time, the upland was reduced to an island chain. The direction of subduction on this and the previous figure have not been clearly established.

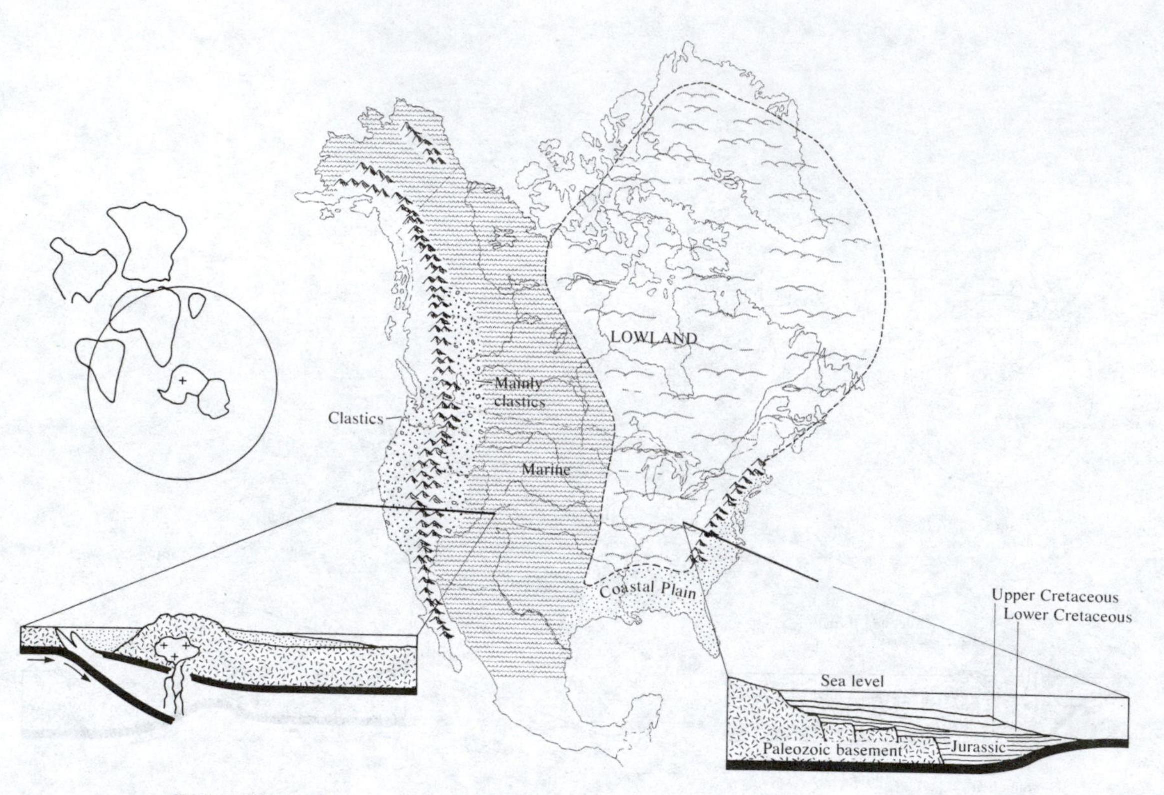

FIGURE 4.25

- Cretaceous paleogeography is shown here. The Jurassic was similar except that the interior sea was less extensive. During the Jurassic and Cretaceous, much of the area west of the western upland apparently was accreted in a number of collisions with volcanic arcs and minicontinents. The trench was near the present coastline, and its volcanic arc was inland at the upland. In California, deformation and batholithic activity were common, especially from late Jurassic to mid-Cretaceous time; this was the Nevadan phase of the Cordilleran orogeny. Along the Gulf and Atlantic coasts, sedimentation began in the Jurassic.

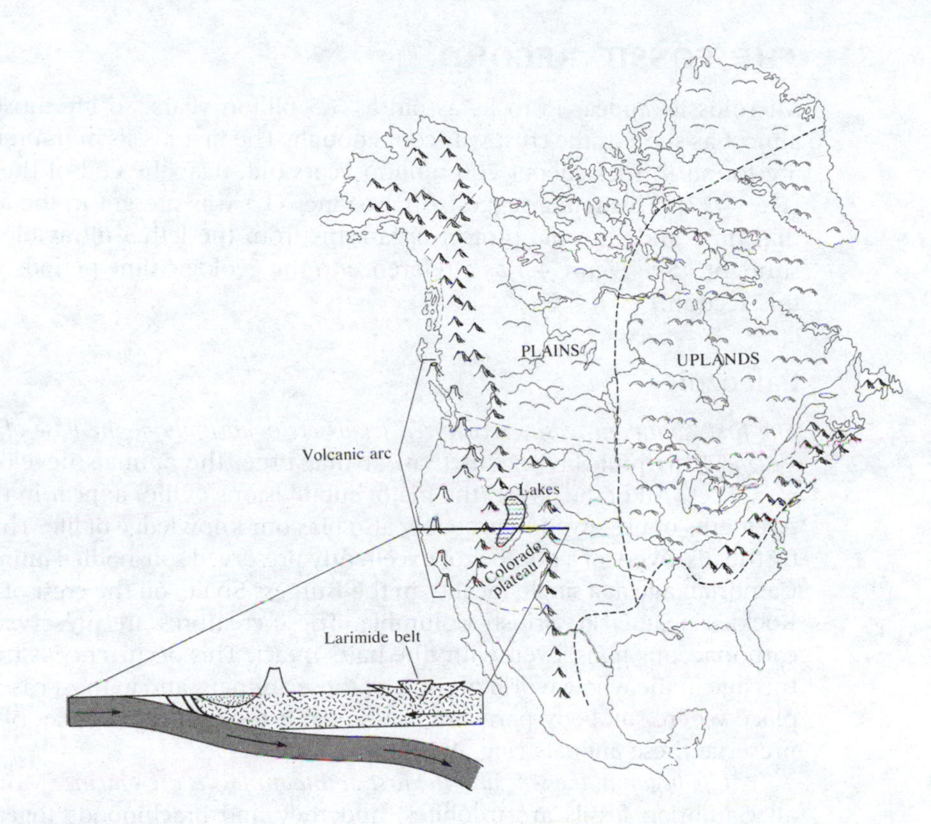

FIGURE 4.26

- In the early Cenozoic, the volcanic arc on the west coast became more active. The rate of movement increased and the angle of downward motion lessened. These changes caused orogenic events much farther inland as well as near the coast. This orogeny is called the Laramide. It was most active in the area of the present Rocky Mountains and occurred from late Cretaceous to early Cenozoic time. It was the last phase of the Cordilleran orogeny. Eastward thrusting occurred from southern California to Alberta, especially in the late Cretaceous, and is called the Sevier phase of the Cordilleran orogeny.

THE FOSSIL RECORD

Microfossils appear in rocks as old as 3.5 billion years, so life must have begun almost as soon as the crust was cool enough. The first fossils of nonmicroscopic life are found in rocks about 700 million years old, near the end of the Precambrian (Fig. 4.27). This fauna suggests that ozone (O_3) was present in the atmosphere at that time and so could protect organisms from the lethal ultraviolet radiation in sunlight. (See Figure 4.3 as a reference to the geologic time periods while reading this section.)

Paleozoic

The first abundant, easily recognized fossils occur abruptly at the base of the Cambrian System. This probably means that, at that time, the animals developed shells or carapaces. All of the phyla (the major subdivisions of life) appear in the Cambrian. Accidents of preservation probably also bias our knowledge of life. This is suggested by the discovery of a number of excellently preserved soft-bodied animals of Middle Cambrian age at a single locality in the Burgess Shale, on the crest of the Canadian Rockies near Field, British Columbia. These creatures are preserved as flattened carbonaceous films, even with fine hairs intact. This occurrence is the only one of this age in the whole world of some of these animals, and in most cases it is the only place where soft body parts are preserved. It took an unusual set of conditions to preserve these animals (Fig. 4.28).

Life began in the sea, and the first abundant fossils are marine. More than half of all Cambrian fossils are trilobites, and they and brachiopods together comprise

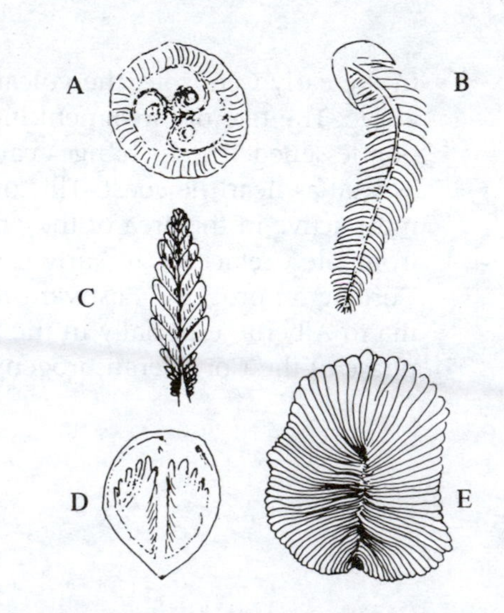

FIGURE 4.27
This unusual Precambrian fauna is found at more than a dozen places. These organisms are found as impressions in sandstone. A. Circular form of unknown affinities, possibly with coiled arms. About 1.9 centimeters (0.75 inch). B. Probable worm. About 2.5 centimeters (1 inch) long. C. Leaflike form of unknown affinities. About 17.8 centimeters (7 inches). D. Oval form of unknown affinities. About 1.9 centimeters (0.75 inch). E. Wormlike form. About 4.5 centimeters (1.75 inches).

FIGURE 4.28
Burgess Shale on Mount Wapta, British Columbia. This Middle Cambrian formation has yielded extremely well-preserved fossils of 130 species. A. Collecting fossils. The first fossils were found by chance in 1909. B. *Waptia fieldensis,* a shrimplike arthropod. C. *Canadia setigera,* a worm. D. *Marrella splendens,* a trilobitelike arthropod. E. *Leanchoila superlata.* F. *Burgessia bella.* (Photos from Smithsonian Institution)

about 90% of Cambrian fossils. Trilobites are good index fossils for the Cambrian. They became less abundant after Early Ordovician time as other invertebrates developed, and the trilobites became extinct near the end of the Paleozoic.

A reconstruction of a Cambrian sea bottom is shown in Figure 4.29. The crowding together of so many organisms shown in this and other dioramas probably never occurred in nature but has been done to show the varieties of life at that time.

FIGURE 4.29
Reconstruction of Middle Cambrian sea floor in western North America. A. Spongelike animal. B. *Marrella,* an arthropod. C. A worm. D. A jellyfish. Several different trilobites are also shown. (From Field Museum of Natural History)

Few predators are found in Cambrian faunas, but they are found in Ordovician and later rocks. Straight-shelled nautiloids probably resembled present-day squids, and some could swim rapidly and catch prey with their tentacles. It may be that shells and skeletons were developed mainly for support in the Cambrian, but evolved into protection in the Ordovician. The development of spiny shells in the late Paleozoic adds credence to this idea.

The marine invertebrates continued to evolve rapidly so that life in the late Paleozoic differed from that of the early Paleozoic. Brachiopods, corals, and bryozoans were abundant, and snails and clams were common. Figure 4.30 shows a late Devonian sea bottom. Also shown in this Devonian scene is the coiled cephalopod, which developed by the coiling of a straight-shelled nautiloid. The coiled cephalopods developed very rapidly and are the most important fossils for dating the late Paleozoic and Mesozoic rocks.

Vertebrates—fish—first appear in the Cambrian. They, too, began in water. The first evidence of their existence is scales and bone fragments found in an upper Cambrian sandstone. Few other fish fossils are found until Late Silurian, although by the Devonian they are dominant. The Devonian is called the "age of fish" because fish apparently ruled the seas.

Plant life on the land also developed rapidly in the late Paleozoic. From a few Ordovician occurrences came the abundant flora of the Devonian. The Early Devonian plants were small and primitive and seem to have grown in marshes or swamps. The plants were successful and evolved rapidly. By Middle Devonian time, forests existed. A lush Pennsylvanian coal swamp is shown in Figure 4.31;

FIGURE 4.30
Late Devonian sea bottom in western New York. A. Bryozoans. B. Crinoid. C. Colonial corals. Several types are shown, and corals are the main element in the fauna. D. Solitary coral. Several types are shown. E. Coiled cephalopod. F. Straight nautiloid. G. Spiny trilobite. Spines probably developed as a defense against fish that had developed by this time. Two smaller trilobites are also in the view. H. Brachiopods. Other types are attached to the coral. J. Sponge to the left of the letter. Also shown are snails. (From Field Museum of Natural History)

these swamps disappeared in the Permian when the climate changed to arid or semiarid as is recorded by the redbed and evaporite deposits and the wind-deposited sandstones.

The first insects appeared in the Silurian, and they became large and relatively abundant during the Pennsylvanian (Fig. 4.31). Scorpions may have been the first animals to breathe air. They have an impervious surface and so were among the first to move into freshwater and finally out of water altogether.

The first land vertebrates developed in the Devonian from the lobe-finned fish. This probably occurred in a freshwater environment because the change from ocean to freshwater solved some of the problems of living out of the water. The body fluids of most animals contain about the same amount and type of salts as seawater. Some way to enclose these fluids to prevent dilution was necessary for the animals to move into freshwater. Once that was accomplished, only a way to get oxygen from the air was necessary. It seems likely that the ability to travel out of the water

FIGURE 4.31
Reconstruction of a Pennsylvanian coal swamp. A. *Lepidodendron.* B. *Sigillaria.* Note the large cockroach. C. *Calamites.* D. Large dragonfly. (From Field Museum of Natural History)

was a great advantage if the pool a fish was in dried out. It could then crawl to the next pool. In the course of moving from pool to pool, it might find something tempting to eat and so began to live more and more out of the water. In this way, perhaps, the first amphibians developed.

The oldest amphibian fossils so far found are in the upper Devonian rocks of Greenland and eastern Canada. They developed from the lobe-finned fish when the fins of these fish developed into clumsy legs. The lobe-finned fish also developed another adaptation that enabled them to live to the next rainy season if their pool dried up. The lungfish can burrow in the mud and breathe air while in a hibernationlike state.

The development of amphibians was a great step toward inhabitation of the continents, but they were still tied to the water. They had to return to the water to lay their eggs. In spite of this, they expanded very rapidly. The reptiles were the next step, and they first appeared in the Pennsylvanian. The reptiles are characterized by the amniote egg that can hatch on land. It is difficult to distinguish between reptiles and amphibians on the basis of skeletons alone because they are so similar. *Seymouria,* found in Lower Permian rocks in Texas, is thought to be a reptile only because fossil eggs are found with it.

Mesozoic

The marine invertebrates of the Mesozoic are very different from those of the late Paleozoic. In the Paleozoic seas, brachiopods were the main bivalves; but in the Mesozoic and Cenozoic, they are largely replaced by clams, including oysters.

Near the end of the Paleozoic, many groups became extinct. No completely satisfactory explanation is available for this great dying. Trilobites and blastoids became extinct. Almost no corals or foraminifers survived, and several types each of bryozoans, brachiopods, and crinoids died out. However, the survivors populated the Mesozoic seas. An interesting point is that some organisms that died out at most places lived on somewhere.

Near the end of the Paleozoic, the continents came together, forming a single supercontinent. Most of the invertebrates that died out about this time lived in shallow seas, and the joining together of the continents would, of course, destroy much of the area of shallow seas. Thus, this major subdivision of geologic eras on the basis of fossils is a natural one. The great change in life was obvious to the early workers who defined the eras.

In the late Paleozoic and the Mesozoic, the cephalopods are almost ideal index fossils. They were abundant, are found in many environments, and they evolved rapidly. The stratigraphy of this interval is based on these animals, which were similar to the present-day nautiloids. They were swimming animals with coiled, chambered shells. They evolved rapidly, with differences between species being the ornamentation of the shell and the shape of the partitions between the chambers (Fig. 4.32). Because they were swimmers, the fossils are found in all environments. It seems likely that they floated after death as well. Why they developed such extremely complex partitions is not known, although the partitions probably strengthened the shell much as corrugations strengthen cardboard. In any case, the cephalopods died out at the end of the Cretaceous except for the nautilus, which has simple, smooth partitions.

The Mesozoic plants, too, were different from those of the Paleozoic, although the change is not apparent until Late Triassic. As mentioned, the lush swamps that produced the Pennsylvanian coal beds disappeared in the Permian as the result of the climatic change to arid or semiarid. These desert conditions, which persisted into Triassic time, resulted in preservation of very few members of the early Mesozoic flora. The few fossil plants found in early Mesozoic rocks, such as horsetails, are holdovers from the Paleozoic. In Late Triassic, the Mesozoic flora of ginkgoes, conifers, ferns, and especially cycads, all of which differed from Paleozoic forms of these plants, was well established. Angiosperms, or flowering plants, appear in early Cretaceous rocks. The angiosperms, which are important elements of present floras, became abundant and spread to all continents in the Middle Cretaceous.

The Mesozoic was the age of the reptiles, and of course the dinosaurs come to mind first. The dinosaurs were not the only reptiles. In the Permian and most of the Triassic periods, the dominant group was the mammal-like reptiles. They became

FIGURE 4.32
Cephalopods. A. *Baculites compressus.* A 75-millimeter (3-inch) fragment of a straight-shelled ammonite. B. Two views of *Ophiceras commune,* a Lower Triassic ceratite from Axel Heiberg Island, Arctic Canada. About natural size. C. Two views of *Scaphites depressus,* Upper Cretaceous with ammonite suture from Alberta. About natural size. (Photo A from Ward's Natural Science Establishment, Inc., Rochester, N.Y.; B and C from Geologic Survey of Canada (109804 and 109804–A, 113737–K and–H).

A

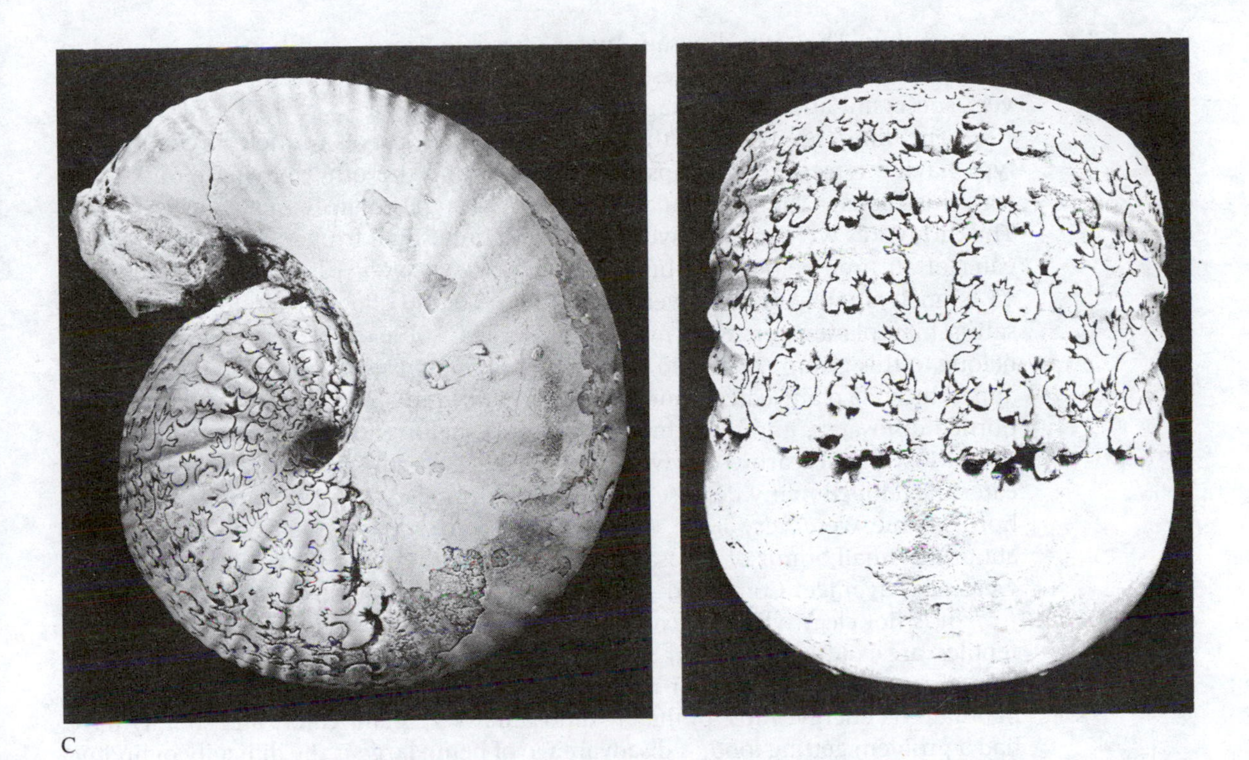

C

greatly reduced in Late Triassic time as the dinosaurs took over. The mammals evolved from the mammal-like reptiles, and the first mammal fossils are found in Upper Triassic rocks. The turtles are found from Early Triassic on. The first fossil lizards are in Upper Triassic rocks, and the first snakes are Cretaceous.

The marine reptiles illustrate how a successful group like the reptiles moved into all environments. They returned to the sea, but they were better adapted than their amphibian ancestors. They developed a system whereby the unhatched eggs were kept in the mother's body. Some Triassic ichthyosaurs have been found with unborn young in the body cavity.

Birds are similar to reptiles in many ways. They differ in having feathers and being warm-blooded. The oldest bird is Jurassic and is known to be a bird rather than a reptile only because three almost complete specimens with feathers have been found in fine-grained limestone. This bird, *Archaeopteryx,* was about the size of a pheasant and had teeth in its beak. Birds are among the rarest of fossils. The first toothed flying reptiles are found in the Triassic rocks. The largest flying reptile so far found is from Texas and has a wing spread of 15.5 meters (50 feet).

The first dinosaurs are found in Upper Triassic rocks. They were small animals, not at all like the huge beasts of Late Jurassic and Cretaceous that were soon to rule the lands. Many of the dinosaurs were small, but it is the big ones that get attention. The term dinosaur means "terrible lizard" and has been applied to two different groups of reptiles. It is an informal but useful term. Dinosaurs were very successful and evolved rapidly.

The difference between the two types of dinosaurs is in their hip bones, one type having typical reptile hips (saurischians) and the other having birdlike hips (ornithischians). The saurischians were of two types, carnivores and herbivores. The carnivores were mainly two legged, with only small front legs; the largest flesh eater of all time belongs to this group: *Tyrannosaurus*—6.1 meters (20 feet) tall; 15.25 meters (50 feet) long; weighing 7300 to 10,000 kilograms (8 to 10 tons). The saurischian plant eaters were mainly four legged; some of the largest animals of all belong to this group *(Apatasaurus,* formerly *Brontosaurus).* The ornithischians were apparently all plant eaters, and the bizarre dinosaurs, such as the duckbills, the horned dinosaurs, and the armored dinosaurs, belong to this group (Fig. 4.33).

Most of the carnivores were two-legged, fast-running predators. The plant eaters developed many defenses. Some could run fast, some had armor, some had horns, some were very large, and some lived in water. Much has been written about the small brains of dinosaurs, but compared with other reptiles, their brains were near average. Compared with mammals, their brains were small.

It is not clear why some of the dinosaurs became so large. It may be because reptiles are cold-blooded and so it is an advantage to be large because a large animal needs a much smaller percentage of its body weight in food each day to maintain its energy. Most of the big dinosaurs were plant eaters and so may have had a problem getting food. A disadvantage of being large is the difficulty of finding shelter in a storm. Size also may have been a defense against predators. On the other hand, it has been suggested that some of the dinosaurs may have been warm blooded, but the evidence is not clear.

The reptiles were very similar on all continents during most of the Mesozoic, suggesting that until the Cretaceous, all continents were close enough to allow migration.

At the end of the Mesozoic, the dinosaurs, almost all of which had specialized to occupy the many different environments, died out. Many reasons have been suggested for the extinction of the dinosaurs. Most of these reasons could account for the demise of single groups or species, but not for all of the diverse dinosaurs. Examples are climatic change, loss of food supply because of development of the angiosperms, disease, and predation of their eggs, perhaps by mammals. Because none of these or the many other suggestions can explain the extinction of all of the dinosaurs, some have looked to extraterrestrial causes. Cosmic rays and solar wind, especially at times of magnetic reversals, have been suggested, but most extinctions here and elsewhere in the geologic record do not generally coincide with times of magnetic reversals. Another possibility is that the impact of a large comet, asteroid, or meteorite could have thrown enough dust into the atmosphere to reduce greatly the amount of the sun's energy reaching the earth's surface. Such a loss of sunlight would destroy some organisms in the food chain and so cause extinctions. In support of this theory is the concentration of elements expected in cosmic dust at

FIGURE 4.33
Cretaceous dinosaurs. On the right, *Trachodon,* a duckbill. Other duckbills are the web-footed *Corythosaurus* in the water, and crested *Parasaurolophus* in the left background. In the foreground is *Ankylosaurus,* an armored dinosaur. In the center background is the ostrichlike *Struthiomimus,* a saurischian dinosaur. (Painting by C. R. Knight, Field Museum of Natural History)

the Mesozoic-Cenozoic boundary at every place where sedimentation was continuous in that time interval.

Cenozoic

In the Cenozoic ocean, the biggest change was the absence of some groups of cephalopods, which were so abundant in the Mesozoic. Clams, oysters, and snails were the main elements of Cenozoic seas, along with echinoids, bryozoans, and foraminifers.

The Cenozoic is the age of mammals. Mammals have many advantages over reptiles. They have a more efficient heart, and hair, making them warm-blooded. Their brain is larger, and the senses, such as smell and especially hearing, are better. Specialized teeth and jaws improve feeding, and the digestive system does not require inactive periods after eating, as reptiles require. Teeth are easily preserved and are excellent mammal fossils. Growth of the young within the mother and a long period of nursing make the young more likely to survive. Just as the reptiles expanded to all habitats in the Mesozoic, so did the mammals in the Cenozoic. Bats are flying mammals; seals, whales, and porpoises are marine mammals.

There are two main types of mammals: the *marsupials,* such as the kangaroo and opossum, that carry their young in an external pouch; and the *placental* mammals that carry their young internally. There is one other type of mammal, monotremes, living in Australia. Only two are known: the duck-billed platypus and the spiny anteater. They have hair and nurse their young, but they lay eggs.

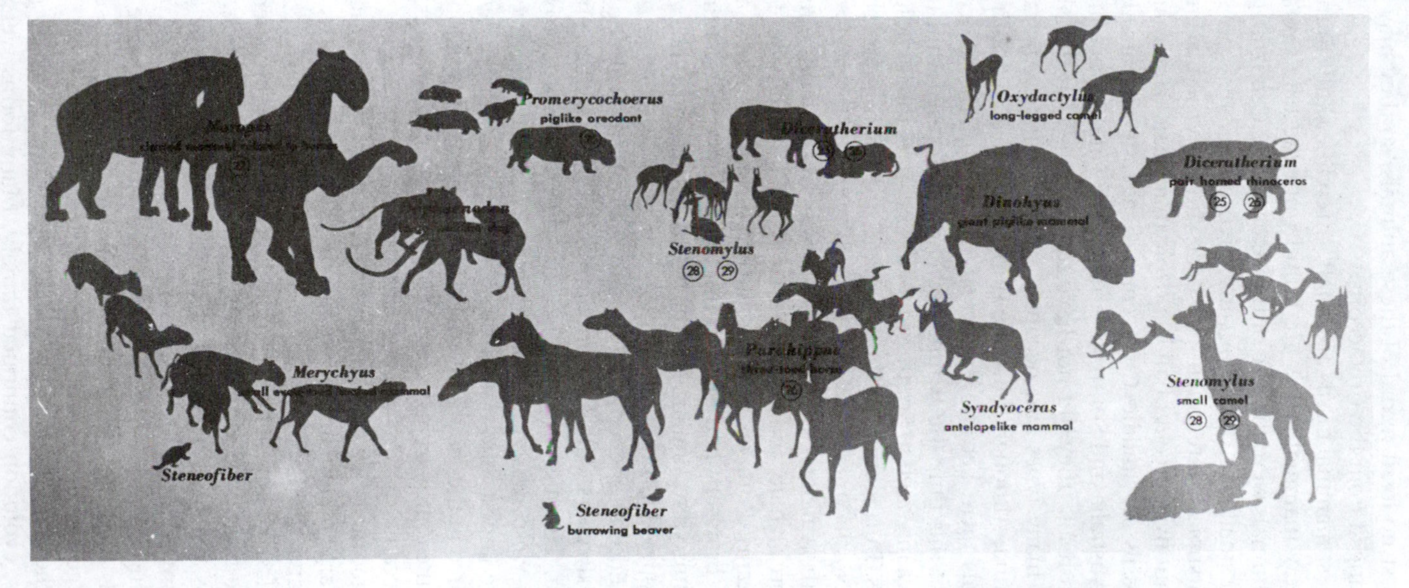

FIGURE 4.34
Lower Miocene of the Harrison Formation of western Nebraska. (Painting by J. H. Matternes, Smithsonian Institution)

There is almost no fossil record of these animals, so they may be "living fossils," almost unchanged from the Cretaceous.

The oldest mammal fossils are small, shrewlike skulls found in Upper Triassic rocks. Very few mammal fossils are found until the Late Cretaceous. The largest Mesozoic mammals were the size of house cats. The Cenozoic placental mammals evolved from small Late Cretaceous insectivores.

The marsupials were much less successful than the placentals. It is estimated that only about five percent of all Cenozoic mammals were marsupials. The oldest marsupial fossils are middle Cretaceous in age. It appears that the marsupials migrated to Australia and South America in the Late Cretaceous and then became isolated. As a result, they did not have to compete with the placentals, and they evolved into many types and occupied many habitats. Australia is noted for its marsupial fauna. Although the same thing happened in South America, many of the marsupials there became extinct at the end of the Pliocene. This occurred because about this time the Isthmus of Panama formed, and the placentals from North America migrated to South America. The marsupials could not compete with the placentals. During most of the Cenozoic, there was some connection between Eurasia and North America.

In North America, the Paleocene was warm and temperate. Paleocene mammals were archaic, and most were small, up to sheep size. They lived in forests and near streams. Primates, rodents, insectivores, carnivores, and browsers were present. The browsers had hoofs and four or five toes. European and North American forms were similar, but the South American types were different. The first horses appeared in the Paleocene.

As the Eocene opened, the climate was subtropical and the animals were largely forest dwellers. All of the modern orders of mammals have been found in Eocene strata, although the species are different from those of the present. Rhinoceroses were present, and in the late Eocene, the deer, pig, and camel appeared. The largest land animal was the rhinoceroslike *Uintatherium*. Toothed whales appeared in the sea, indicating that this habitat was soon occupied after the extinction of the marine reptiles. In late Eocene time, more grasslands appeared as the climate became somewhat cooler and drier.

The Oligocene had a mild, temperate climate in North America, and the Great Plains was a large floodplain with many rivers and some forests. The archaic forms disappeared and the fauna took on a more modern aspect. The drier climate favored the faster, long-legged, hoofed grazers and browsers. Cats and dogs appeared. The titanotheres, which had begun in early Eocene as small rhinoceroslike animals, reached the size of elephants in the Oligocene and then died out. *Brontotherium* was the largest mammal ever to live in North America.

The drying and cooling trend continued into the Miocene in North America, and grasslands expanded, favoring rapid evolution of grazers such as horses (Fig. 4.34). The large cats, bears, and weasels also appeared.

In the Pliocene, the climate cooled, remaining relatively dry. As a result of these changes, evolution continued rapidly. Many forms were weeded out, and those remaining became more specialized. The trend continued into the Pleistocene and produced our present flora and fauna.

The Pleistocene life of North America was dominated by large animals such as mastodons, mammoths, ground sloths, saber-toothed cats, bears, and giant beavers, as well as horses and camels. This was a time of great climatic change everywhere. Most of the animals just listed died out in North America about 8000 years ago, when the last glacial stage was retreating. No one knows why this great extinction occurred, although hunting by early humans has been suggested. It was not the climatic change because they lived through the glacial advances.

If, after reading this book, you are unable to look at a countryside or an outcrop without wondering what it is, how it formed, or what it means, then this book will have served its purpose.

KEY TERMS

Relative date
Superposition
Cross-cutting
Absolute date
Geologic time scale
System
Period
Unconformity
Angular unconformity
Correlation
Index fossil
Radioactive date
Half-life

QUESTIONS

1. How does one interpret the depositional environment of a layer of ancient rocks?
2. Describe in general terms how the geologic column was established and discuss whether it consists of natural subdivisions in North America.
3. What are unconformities? Describe the various types.
4. How was the value of fossils in determining geologic age established?
5. What is meant by correlation?
6. List and briefly describe the methods of correlating rocks.
7. Why is it generally easier to date marine rocks than continental sediments?
8. How are metamorphic rocks dated? What dates are required?
9. Which method of radioactive dating is best and why?
10. What are index fossils?
11. What effect would a small amount of modern root material have if mixed with a sample to be dated by the carbon-14 method?
12. How can an angular unconformity be distinguished from a low-angle thrust fault?
13. Fossils aid in determining the age of a rock unit. How else do they aid in geologic interpretation?
14. Prove to your friend, who is an art major and has not taken any science, that the earth is very old.
15. Distinguish between relative and absolute dating in geology.
16. Explain why the earth melted soon after its formation.
17. How did the earth's layered structure form?
18. When did the first life form?

19. Can the beginning of Cambrian time be recognized at all places where sedimentary rocks of this age occur?
20. What is the age of the oldest rocks so far found?
21. Where are Precambrian rocks exposed on North America?
22. What are some of the explanations for the sudden appearance of abundant fossils at the base of the Cambrian?
23. Why is Precambrian history so difficult to interpret?
24. List as many of the assumptions made in interpreting geologic history as you can.
25. In what continent was the "south" pole in early Paleozoic time?
26. Where was the equator in North America in early Paleozoic time?
27. What is the importance of the Burgess Shale fauna?
28. What are typical Cambrian fossils?
29. What were the first vertebrates, and when did they appear?
30. When did the first land plants appear?
31. When did the orogeny that produced the Appalachian Mountains occur?
32. Which of the present continents were in contact at the end of the Paleozoic?
33. When did the amphibians first appear? From what did they evolve?
34. When did the reptiles first appear?
35. What are some differences between reptiles and amphibians?
36. When did the first insects appear?
37. Name some organisms that became extinct near the end of the Paleozoic.
38. Which continents drifted apart during the Mesozoic?
39. When did the first flying reptiles appear? The first birds?
40. What was the time of the dinosaurs?
41. Describe the various types of dinosaurs and other reptiles of the Mesozoic. Which ecologic areas or niches did they occupy?
42. How does the marine life of the Cenozoic differ from that of the Mesozoic?
43. What are the two main types of mammals?
44. Describe the changing climates of the Cenozoic and how these affected the development of mammals.

SUPPLEMENTARY READING

General

Nichols, R. L., "The Comprehension of Geologic Time." *Journal of Geological Education* 22(2) (March 1974): 65–68.

Raup, D. M. "Biological Extinction in Earth History." *Science* 231(4745) (March 28, 1986): 1528–33.

Precambrian

Calvin, Melvin. "Chemical Evolution." *American Scientist* 63(2) (March–April 1975): 169–77.

Cloud, Preston, James Wright, and Lynn Gover, III. "Traces of Animal Life from 620-million-year-old Rocks in North Carolina." *American Scientist* 64(4) (July–August 1976): 396–406.

Conway, Morris, S. "The Search for the Precambrian-Cambrian Boundary." *American Scientist* 75(2) (March–April 1987): 156–67.

Glaessner, M. F. *The Dawn of Animal Life.* Cambridge: Cambridge Univ. Press, 1984, 244 pp.

Goodwin, A. M. "Precambrian Perspectives." *Science* 213(4503) (July 3, 1981): 55–61.

Head, J. W., and others. "Geologic Evolution of the Terrestrial Planets." *American Scientist* 65(1) (January–February 1977): 21–29.

Kerr, R. A. "Origin of Life: New Ingredients Suggested." *Science* 210(4465) (October 3, 1980): 42–43.

King, P. B. *Precambrian Geology of the United States: An Explanatory Text to Accompany the Geologic Map of the United States.* U.S. Geological Survey Professional Paper 902. Washington: U.S. Government Printing Office, 1977, 85 pp.

Paleozoic

Bambach, R. K., C. R. Scotese, and A. M. Ziegler. "Before Pangea: The Geographies of the Paleozoic World." *American Scientist* 68(1) (January–February 1980): 26–38.

Crawford, T. J., and others. *The Structure, Stratigraphy, Tectonostratigraphy, and Evolution of the Southern Part of the Appalachian Orogen.* U.S. Geological Survey Professional Paper 1475. Washington: U.S. Government Printing Office, 1988, 173 pp.

Palmer, A. R. "Search for the Cambrian World." *American Scientist* 62(2) (March–April 1974): 216–24.

Rudwick, M. J. S. *The Great Devonian Controversy.* Chicago: Univ. of Chicago Press, 1985, 494 pp.

Whittington, H. B. *The Burgess Shale.* New Haven: Yale Univ. Press, 1985, 151 pp.

Mesozoic

Alvarez, L. W., and others. "Extraterrestrial Cause for the Cretaceous-Tertiary Extinction." *Science* 208(4448) (June 6, 1980): 1095–1108.

Bakker, R. T. "Dinosaur Renaissance." *Scientific American* 232(4) (April 1975): 58–78.

Bakker, R. T. *The Dinosaur Heresies.* New York: William Morrow & Co., 1986, 481 pp.

Buffetaut, Eric. "The Evolution of the Crocodilians." *Scientific American* 241(4) (October 1979): 130–44.

Desmond, A. J. *The Hot-Blooded Dinosaurs: A Revolution in Paleontology.* New York: Dial Press, 1975, 238 pp.

Howell, D. G. "Terranes." *Scientific American* 251(5) (November 1985): 116–25.

Langston, Wann, Jr. "Pterosaurs." *Scientific American* 244(2) (February 1981): 122–36.

Mash, Robert. *How to Keep Dinosaurs—The Complete Guide to Bringing Up Your Beast.* New York: Viking/Penguin, 1983, 72 pp.

Russell, D. A. "The Mass Extinctions of the Late Mesozoic." *Scientific American* 246(1) (January 1982): 58–65.

Valentine, J. W., and E. M. Moores. "Plate Tectonics and the History of Life in the Oceans." *Scientific American* 230(4) (April 1974): 80–89.

Cenozoic

Krantz, G. S. "Human Activities and Megafaunal Extinctions." *American Scientist* 58(2) (March–April 1970): 164–70.

Molnar, Peter. "The Geologic History and Structure of The Himalaya." *American Scientist* 74(2) (March–April 1986): 144–54.

Trimble, D. E. *The Geologic Story of the Great Plains.* U.S. Geological Survey Bulletin 1493. Washington: U.S. Government Printing Office, 1980, 55 pp.

Wolfe, J. A. "A Paleobotanical Interpretation of Tertiary Climates in the Northern Hemisphere." *American Scientist* 66(6) (November–December 1978): 694–703.

Index